James Gregory

Annual Circular and Retail Catalogue of Warranted

Vegetable and Flower Seeds

January, 1880

James Gregory

Annual Circular and Retail Catalogue of Warranted Vegetable and Flower Seeds
January, 1880

ISBN/EAN: 9783337372378

Printed in Europe, USA, Canada, Australia, Japan

Cover: Foto ©berggeist007 / pixelio.de

More available books at **www.hansebooks.com**

JANUARY, 1880.

ANNUAL CIRCULAR

AND

RETAIL CATALOGUE

OF

WARRANTED

Vegetable and Flower Seeds.

GROWN AND SOLD BY

JAMES J. H. GREGORY,

MARBLEHEAD, MASS.

Catalogues Free to All.

☞ **PLEASE NOTICE.** My customers will please notice that this season, for the first time, that I have given my rates for quarter-pounds, pounds, pecks and bushels *in the body of my catalogue in connection with the ounce and package quotations,* and not on two separate pages as formerly. I have done this for the convenience of my customers and to gain some much needed space for the catalogue. My prices are by mail postpaid by me, or by express or freight at purchaser's expense. When not ordered by mail it is desirable that customers should indicate whether they wish their seed sent by express or freight ; when no preference is given I will use my best judgment in their behalf. I make no charge for boxes used in packing. ☜

☞ When comparing my rates with other dealers, my customers will please remember :—1st, *That all orders, (with the exception of Peas and Beans, by the peck and bushel, and Danvers Onion seed) to the amount of five dollars and upwards are subject to a cash discount as stated below* ; 2d, *That I grow over half the varieties of the seed I sell* ; 3d, *My three warrants.* ☜

JAMES J. H. GREGORY'S
Annual Seed Catalogue for 1880.

The past season has been a very unfavorable one for the ripening of Beans and Peas. Onion seed is also more scarce than usual, and this is especially true of the Early Danvers variety, of which, it is safe to say, there is not enough of good seed to supply the average demand. My own crop is short and I would advise my customers to order *early.*

☞ My seed growing facilities now include five farms, located in three towns, carried on directly under my own supervision, with about one hundred and twenty-five acres in annual tillage. With these advantages for complete isolation I grow about a hundred and fifty varieties of the vegetable seed contained in this catalogue. *..⁴* All roots, bulbs, &c., intended for seed purposes, are selected with the utmost care.

THE THREE WARRANTS. All seed sent out from my establishment are sold under three warrants; viz.: 1st, That my seed shall be what it purports to be, *so far as that* I hold myself ready to refill the order anew, gratis, in other seed, should it prove defective in any respect. 2d, That all money sent for seed shall reach me, with the single proviso, that *all sums to the amount of one dollar and upwards* be sent in the form of a Post Office Order on Marblehead, Registered Letter, Draft, or Cashier's Check. 3d, That the seed ordered shall reach every one of my customers. Thus I warrant everything but the crop.

There happens occasionally an instance, where I have complaint against the vitality of seed, in which the seed was grown by myself that season, and thousands of customers were supplied *from the same lot* to their satisfaction. Such cases are very obviously of the class where the cause of the trouble was owing to the season, or soil, or manner of planting ; in short to some cause utterly beyond my control and consequent responsibility. It is my labor and my anxiety to send out none but *just such seed as I would be willing to plant myself,* and the thousands of cheering letters that I open are very pleasant testimony to the success of my undertaking With all care that it is possible to exercise, mistakes will occasionally, (though very rarely), occur ; these my customers will always find me ready to rectify in accordance with the promise of my three warrants.

PAYMENT FOR SEED. All sums to the amount of one dollar or upwards should be sent in Cashier's Checks on New York or Boston, Drafts, Money Orders on Marblehead or in Registered Letters. When the value of seed ordered is less than one dollar, the sum sent is at my own risk, and when the value is greater than one dollar, the amount remitted will also be at my risk, *provided* it be sent by Post Office Order, Draft, Cashier's Check, or Registered Letter. When in making change it is necessary to enclose Postage Stamps, *be sure not to moisten them in the least* ; the higher denominations, such as tens and upwards, would be preferred. Money Orders, when of the value of one dollar and upwards, may be purchased at my expense, the cost being deducted from the amount remitted. If Money Orders are not for sale at your office, they can oftentimes be procured at a neighboring town. *Cash must accompany all orders.* When necessary to send specie, wrap it securely in a bit of paper, to prevent it from getting loose and cutting through the envelope.

I would advise my customers *not* to order their seed by Express C. O. D., particularly when the orders are small, as I have to add the cost of collection and return charges to the cost of the seed to make myself whole.

ADVANTAGES OF BUYING SEED DIRECTLY FROM THE GROWER.— But few seed dealers grow any of the seed they sell,—the business of seed growing and that of seed selling being almost entirely distinct. Hence, as a general rule, seed dealers *know* only what is *told* them of the careful selection of seed stock and of the freshness and purity of their seed ; now if the person of whom they purchase should be careless, ignorant or dishonest, you who plant have to suffer, as the dealer can only reaffirm what is told him. On the other hand, if the seed dealer grows his own seed, he is able to affirm what *he himself knows* as to its freshness and purity ; and thus you who purchase have the invaluable guarantee from his own knowledge. *It is that I may be able to give this guarantee that I raise so many varieties, some of them at double the cost at which I could purchase them.* The public will thus understand how greatly it is for their advantage to sustain me in this effort to combine the business of seed grower and seed dealer.

From what I have said let it not be inferred that I raise all the varieties of seed I sell ; I do not ; many choice varieties I import from England, France and Germany, some of which cannot be raised in this latitude, others I purchase of reliable dealers, or have grown for me by careful men, whom I supply with stock seed of my own raising.

☞ When writing *that seed have failed to reach you, always* REPEAT YOUR FIRST ORDER, or send for the same value in other kinds of seed.

☞ BE CAREFUL TO WRITE YOUR ADDRESS IN FULL, giving the STATE as well as the town, *for a town of the same name will oftentimes be found in a dozen States. I find by my record that in one season 218 of my customers forgot to sign their names to their letters.* Before closing your orders be sure that you have given your address in full, your name, especially, being written very distinctly. *Please be particular in this matter.*

THE POSTAGE LAW. It matters not how many varieties are sent in the same package, provided the weight of the entire lot does not exceed four pounds. If a larger lot is wanted, it may be divided into lots of four pounds each. This law, in effect, *brings my seed establishment to every man's door.*

Large Orders from Market Gardeners, Dealers, Clubs, and Granges of Husbandry.

Five per cent. discount allowed on all orders for seed to the amount of five dollars and upwards, (excepting Peas and Beans by the peck and bushel, and Onion seed, for discount on which latter see page 32) whether the order be for packages, ounces, pounds or quarts, either separately or combined ; eight per cent. when the amount ordered is as high as ten dollars and under twenty-five ; ten per cent. when twenty-five and under forty dollars ; and twelve and a half on orders to the amount of forty dollars and upwards. For special discounts on Flower seeds see page 53. Terms to Dealers, Clubs and Granges of Husbandry on application. I do not send out seed to be sold on commission.

Early Bleichfield Cabbage.

This new Cabbage is well worthy the attention of farmers and market gardeners. It is of German origin and comes highly recommended. I raised it last season on a large scale as an improvement on Early Quintal, and am much pleased with it. I find it to be the earliest of the large hard heading drumheads, maturing earlier than the Fottler's Brunswick. The heads are large, very solid, tender when cooked and of excellent flavor. Stump short. It is as reliable for heading as any cabbage I have ever grown. The above engraving I have had made from a photograph of a specimen grown on my grounds. The Bleichfield appears to hold a place distinctly by itself among the early drumheads raised in the United States. Price per package, 10 cts; per ounce, 50 cents; per pound, by mail, $5.00.

Gerry Island Cauliflower.

I would advise my customers to test this Cauliflower for earliness side by side with any variety in the market. If it deports itself in other localities as it has in my experimental ground for the past two seasons, it will excel all other varieties in earliness, while in reliability for heading it will be excelled by none. Per package, 30 cents ; per ounce, $4.00.

Chinese Bean.

Among the various vegetable seed sent over here to our Centennial by the inhabitants of the "Great Flowery Kingdom," as our brethren of the vast Chinese Empire love to call themselves, there were two or three varieties, the merits of which I find are worthy to be examined into by us "Outside Barbarians." One of them is a singular species of bean, of which the above is an engraving which I have had copied from a photograph of an average sample of a lot grown on one of my farms. I counted eighty pods on the vine containing from six to ten beans each. The beans themselves are quite small, of a light drab color, and in shape half way between a pea and a bean. Though the individual beans are quite small, (I think in the engraving they appear to be larger than they really are) yet the vines crop so enormously they yielded with me at the rate of forty bushels to the acre. In quality this bean reminds me of the "Calavanser." What place it may occupy in the economy of the agriculture of the country is yet a matter of experiment. Per package, 15 cents.

Bay View Melon.

This new sort received first prize at the annual exhibition of the Pennsylvania State Agricultural Society, as a new variety superior to the old sort. Green fleshed, sweet and spicy ; it has been grown to weigh seventeen pounds. It is hardy, very vigorous and productive. Ripens a few days after

Jenny Lind. Per ounce, 30 cts.; per package 15 cts.

ROSE BEAN.

A new bush bean, remarkably prolific and vigorous. It is exceedingly productive and the beans are exceptionally large. The vines are the stoutest and the leaves the largest of any bean I have ever raised. One of my workmen reports an extraordinary yield of the new bean in his garden. The color of the bean is of a rich dark rose. It is entirely distinct from the variety known as the Chili or Red Kidney. Per package, 15 cents ; per quart, 45 cents by express, 75 cents by mail.

TOMATO, TURK'S TURBAN, SCARLET.

This new Tomato is particularly early and prolific, producing rarely less than ted to fifteen fruits in a cluster. A great peculiarity of the sort is that each fruit contains only a few seeds and these lie close to the outer surface, so that the whole of the rest of the fruit, and notably the centre, is fleshy and in consequence fit for use. It is of delicious flavour, and though small in size, excellent for preserving. The plants are small-foliaged and of robust growth. In a bearing state they are very showy, and may be utilized for decorative purposes. Per package, 20 cents.

ALPHA TOMATO.

Every season I plant many new varieties of tomato advertised, with many new sorts or strains that are not advertised, sent me for trial by my friends in various parts of the United States and Canada. These are planted in my experimental grounds, side by side with standard sorts, under precisely the same conditions, my object being to determine whether or not any of the new sorts are superior to the old. Occasionally a new comer will exceed the old standards in some one characteristic, taken all in all be no better, if as good as they are. Still more rarely a new variety will be developed, which taken as a whole (and earliness counts very high) will really prove to be an acquisition. This season in a trial test, which was every

way fair, I found one real acquisition in the "Alpha."

ALPHA TOMATO.

Passing over the experimental plot, note book in hand, I was most agreeably surprised to find one variety decidedly ahead of the others in ripeness. Turning to my note book I found that this was the "Alpha." This new tomato is of a roundish shape, smooth and symmetrical. It grows to a good market size, ripens capitally around the stem, and does not crack when, after a rain, some other kinds shows a decided weakness in that direction. This tomato is solid, rich colored, very symmetrical, and taken altogether, about as handsome a variety as is carried to market. My customers could not do better than try the new comer on their grounds, for if it does as well with them as it has with me, they will find they have a most decided acquisition. Per package, 20 cents ; per oz., $1.50. Extra selection, per package, 25 cents ; per oz , $2.00.

WHITE GERMAN CUCUMBER.

As will be seen by the engraving, this is an enormously large and exceedingly handsome cucumber, indeed, some of my workmen vote it to be the handsomest variety we have ever grown. The white color is peculiarly clear and strong. It surpasses most of the foreign varieties in vigor and productiveness. To those who have the English fancy for a cucumber having but few seed, and to all who desire to raise the handsomest possible variety for exhibition purposes, I know of no sort which can give greater satisfaction than the White German. Per package, 25 cents.

ESSEX HYBRID SQUASH.

This is a cross between the Turban and the Hubbard, having the shape of the former and the shell of the latter. It appears to combine the qualities of the two parents. The characteristics are said to be permanent and the squash itself to be earlier than either the Hubbard or Turban. Should this be so it will be a decided acquisition to our list of squashes. Not having as yet cultivated this variety, I am not able to speak from experience. Per package, 15 cents ; per oz., 40 cents; per lb., $3.00.

MARBLEHEAD EARLY SWEET CORN.

Last season I raised thirty-eight varieties of corn, including all the early sweet varieties, my principal object being to compare them for earliness with a new variety which I grew for the first time the season previous. To my surprise and great pleasure the new variety, when tested side by side with such standards as Early Minnesota and Narragansett, proved to be *a week earlier than any of them.* A similar test this season with the Tom Thumb, Dolly Dutton, and other early varieties, gave the same result. I have named this new variety the "Marblehead Early." In all its characteristics except earliness it bears a close resemblance to the Narragansett. The stalk is dwarf in its habit of growth, and sets its ears very low down. I send out the Marblehead Early Corn as *the earliest variety of sweet corn* cultivated. It is of fair market size and very sweet. For prices see page 22.

EARLY AMBER SUGAR CANE.

This new variety of sugar cane is making quite a sensation among the farmers of Minnesota, from the perfect success which they have met with in the manufacture of both syrup and sugar. Over 200,000 gallons of syrup were made from it in Minnesota in 1878, and this season a single firm has sent to market 43,000 lbs. of sugar. It has been experimented with in Massachusetts on a large scale, and the results have been most satisfactory; so I think it may now safely be said that after many years experiment and hosts of failures, with several varieties of Sorghum or Imphee, in the Early Amber, which is the most improved form of Otaheitan, a grand success has been reached at last. Having been grown with such success in Lat. 44.30, it can be successfully raised in almost every portion of the northern states. The yield per acre of syrup is from 140 to 280 gallons, and the produce of sugar is about 6 lbs. to the gallon of syrup. My seed has been specially selected for purity.

Says the St. Paul Pioneer: Mr. Miller actually raised 900 pounds of excellent cooking sugar from each acre of ground. Besides this there is left some five or six pounds of syrup from each gallon. All that we have seen is thick with sugar, and sells for 90 cents per gallon. The cultivation required is the same as for corn.

☞ WITH EVERY PACKAGE OF SEED I WILL SEND A SAMPLE OF THE SUGAR MADE FROM THIS NEW SUGAR CANE. My seed is selected with special care for purity.

Per lb. by mail, postage paid by me, 40 cents; per qr. lb., 15 cents; per package, 10 cents. The standard work of Mr. I. A. Hodge, giving full instructions for the manufacture of sugar, sent to any address for $1.00.

CRAWFORD'S HALF DWARF CELERY.

This variety is very extensively grown by the market gardeners who supply the markets of New York City. It has a rich, nutty flavor and great vigor of growth, and with those who do not succeed with the Boston market has grown to be quite a favorite. Per lb., by mail, $4; by express, $3.85; per oz., 40 cents; per package, 10 cts.

FERRY'S PEERLESS WATERMELON.

Of medium size, thin rind, mottled green, flesh bright scarlet, solid to centre, very sweet, very prolific. An excellent sort for garden cultivation. Per package, 6 cts.; per oz., 15 cts; per lb., by mail, $1.25; by express, $1.10.

WHITE RUSSIAN SPRING WHEAT.

The White Russian Wheat is a bald white chaff wheat, of a much lighter red color than most varieties of spring wheat, and has proved itself to be the best wheat ever grown in Wisconsin. No wheat ever tried in this country has received a more unanimous commendation from those who tried it. The White Russian Wheat has astonished all who sowed it. Such long, strong, healthy yellow straw, standing straight several days after ripening, and bearing large long white chaf heads, well filled with plump kernels, weighing oftentimes from 60 to 62 lbs. to the measured bushel, while the wheat produces 5 to 10 bushels more per acre than other once well thought of varieties.

The White Russian Wheat stands well after being ripe, and is not liable to lodge or rust when green; it is decidedly healthy and has in many cases produced a full average crop, where other varieties along side of it have failed.

The following statements were received from parties who tried the wheat last year.

From Henry Cameron, Pilot Grove, Grayson Co. Texas. "I purchased and sowed 4 lbs. of the White Russian Wheat last March, from which I threshed two bushels. There were heads in my patch eight inches long. The grain is very fine and large. My neighbors would like to buy."

From A. G. Quin, Humphrey, Plat. Co. Neb. "Of the White Russian Wheat I sowed 1-4 of an acre, and harvested 11 1-2 bus. at the rate of 46 bus. per acre. I shall sow all I raised."

From Myron Turrell, Bay View, Wis. "The twelve bushels of White Russian Wheat I purchased was divided among five of my neighbors, reserving 3 bushels for myself, from which I harvested 80 bushels of No. 1 wheat, or 40 bushels per acre. My neighbors had equally good crops as mine. It is the best wheat ever raised in this vicinity, and yielded twice as much as fife wheat."

From George W. Abbott, Harvard, Clay Co., Nebraska, Nov. 15th, 1877:—I sowed one bushel of White Russia Wheat from which I threshed 49 3-4 bushels of very good wheat. Other wheats in this vicinity yielded 10 to 20 bushels per acre.

From Jas. M. Williams, Monroe, Iowa, Oct. 12th, 1877:—I sowed two bushels on one acre and harvested 42 1-2 bushels. I sowed two other varieties, and I think the White Russian produced double that of any sown on my farm.

From Daniel Pulver, Delhi, Iowa, Oct. 30th, 1877:—I sowed two bushels of the White Russian Wheat by the side of two bushels of the "Last Nation" Wheat, which I sent to Minnesota for, and find in comparing them, that the White Russian is far in advance, both in quality and quantity, besides it is about a week earlier and does not lodge down.

From Ed. Mayou, Stewartville, Minn., Oct. 13th, 1877:—The half bushel of White Russian Wheat I sowed on half an acre, from which I harvested 22 bushels of good plump wheat.

From J. E. Johnston, Des Moines, Iowa, Oct. 14, 1877:—The two bushels of White Russian Wheat I sowed on a little more than one and a fourth acres, and harvested 41 bushels of nice plump wheat. I think it will yield 40 bushels to the acre in a good wheat season. It beats all other spring wheats in this vicinity. I will save all I raised for seed. I wish I had sent for 10 bushels last spring, which would have furnished me with enough seed for my next year's crop.

Prices.—3 lbs by Mail 90 cts.; half bushel, $1.50; 1 bushel including bag $2.65; 10 bushels including bags, $2.45 per bushel.

TEOSINTE (Reana luxurians.)

This gigantic grammæa is perennial in hot climates. It somewhat resembles Indian Corn in aspect and habit of growth, but the leaves are much longer and broader and the stock is filled with sweeter sap, and is likely to prove of value for the production of sugar. In its perfection it produces a great number of shoots, growing three or four yards high, very thickly covered with leaves, yielding such an abundance of forage that one plant is estimated to be sufficient to feed a pair of cattle for twenty-four hours. In the extreme South, Teosinte would be a perennial. In the north a single seed will make from 12 to 16 stalks, when planted in the open ground, and from 25 to 30 if first started in a hot bed, attaining to the height of five and six feet with a vast mass of long broad leaves.

Mr. S. A. Cook, of Georgia, who grew Teosinte last season, writes me as follows: "It surpasses either Corn or Sorghum as a soiling or fodder plant. I counted 85 stalks from one seed. They grew eleven feet in height, and but for excessive dry weather would have been much taller. Cows are extravagantly fond of them. Per ounce, 50 cents; per package, 15 cents.

THE WHITE EGG TURNIP.

Last season I experimented with every variety of the early turnips found in the catalogue of the most extensive seed grower of England (twenty-three in number) to determine whether this turnip claimed to be a new American variety really was such. After a careful comparison with all these foreign varieties I am satisfied that it is a new variety, as distinct from any of the sorts tested as they are from each other. The engraving gives a very correct idea of its shape. It is perfectly smooth, of a pure clear white, growing half out of ground, and at times to the size of a ruta baga, a very choice kind for table use, of excellent flavor, sweet and mild. It pulls clean from the ground, and with its thin, snow white skin looks almost as attractive as a basket of eggs. It is a first rate keeper for winter use. In both appearance and quality it holds a very high rank and must be considered as quite an addition to our Fall Turnips.

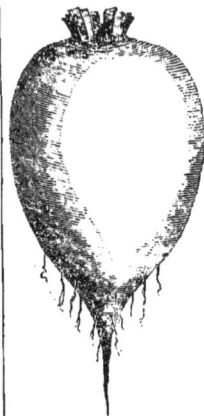

Comes to size for use just after Early Red Top. Per lb. by mail $1.00; half pound, 60 cts.; per oz., 15 cts.; package, 10 cts.

EGYPTIAN SWEET CORN.

I find after testing this new corn on a large scale that it is deserving every word of praise Mr. Hyde has bestowed upon it.

I don't think I ever ate a white variety of sweet corn so sweet and so tender as this. I present herewith an engraving made from a photograph of stalk grown in my garden. It is very tall and very late, coming in about the time of Stowell's Evergreen, but surpasses that variety in sweetness.

Mr. Hyde, who introduced this variety, says—" The ears grow very large and very prolific. Last season and this I put the price down to $3 per hundred wholesale, and could not supply the demand, and I have orders now for next summer for all that I can sell in a green state at $3.00 per hundred. I received first premium for this corn at all County Fairs that I sent it to for exhibition in this state. It has a sweet, rich, delicious flavor that I have never found in any other variety, and is exceedingly tender. For canning I think it far superior to any other variety, having sold all I put up at $2.50 a dozen cans, which is more than any other canned corn will bring in America." For price, see page 22.

My customers writes me of this new corn as follows:

Mr. E. R. Ogier of Maine, writes :—" My Egyptian Sweet Corn had from three to five ears on a stalk, and was the sweetest corn I ever tasted."

Mr. Hardin G. Back of Mass., writes :—" It was the tenderest and sweetest corn I ever saw; and as for yield I never saw anything beat it. I have now what grew from one kernel, and there are nine ears on the main stock and six on the suckers, making fifteen in all, seven of them well filled out. I keep it for a show."

"The yield was enormous, giving me on an average three and in many instances five *perfectly* developed ears to a stalk! I must say that it is the sweetest and by far the *tenderest* corn I ever saw which is also the testimony of my neighbors who tried it. Think you have underrated it in your catalogue."   S. F. Coombs, *Mass.*

PRICKLY COMFREY.

This new forage plant is extensively grown in Europe for the feeding of stock. It is a deep rooted plant, and

even in the hottest seasons will yield several cuttings of forage. It comes in earlier and lasts longer than almost any forage crop. The method of propagation is by roots only. The cultivation is very simple. In well ploughed and well manured ground plant the cuttings three feet apart each way, giving them a liberal dressing of manure the first winter, and no further expense is needed. Cuttings by mail, 40 cents a dozen ; $2.50 a hundred. Rooted plants by express or freight, $1.25 per lb.

The Butman Squash.

In the essay which received the premium of the Massachusetts Horticultural Society, the Butman for quality, is put at the head of all varieties of winter squashes.

This new squash is the only one of our running varieties *known to have originated in the United States*. Externally, it is very distinct in color from any other kind in existence, being a bright grass green intermixed with white. In size and productiveness it resembles the Hubbard ; it has a thick shell and is thick meated. The color of the flesh is quite striking, being of a lemon color. It is exceptionally fine-grained, in this respect surpassing every other variety, and is very smooth to the palate. It is remarkably dry, sweet and delicious, with a flavor different from the Hubbard, Marblehead or Turban, being entirely free from the pumpkin-like flavor occasionally found in the Hubbard. I am inclined to the opinion that the period when the Butman Squash is in its prime is from October to January, though as a keeper it is equal to the Hubbard. For price, see page 29.

[From Editor of American Agriculturist.]
NEW YORK, December, 1874.

MR. GREGORY,

Dear Sir :—The Butman Squash was duly received and has been tested. I can say no more than that it seems to me that every good quality of every good squash is in this, concentrated and combined. When you get any better squash, please send it to
Truly Yours,   GEORGE THURBER.

Cocoanut Squash.

A magnificent little squash for table use, very prolific, yielding from six to a dozen to the vine. In beauty it excels every variety of the Squash family ; indeed, wherever grown, specimens very naturally find a place on the mantel piece as ornaments to the parlor—not being surpassed in beauty by any of the gourd family, the color is an admixture of cream and orange, the latter color predominating in the depressions between the ribs ; while the bottom over a circle of two or three inches in diameter is of a rich grass green. The flesh is fine-grained, very solid, (the squash must be remarkably heavy for its size), and the quality excellent, closely resembling Canada Crookneck, but in every way much superior. Per pkg. 10 cts.; per oz. 30 cts.; per lb. by mail, $2.50.

The Marblehead Squash.

This squash, as a rule, is characterized by a shell of a more flinty hardness than the Hubbard. It is usually thicker and flatter at the top. It has a greater specific gravity. The flesh is of rather a lighter color than the Hubbard, while its combination, in good specimens, of sweetness, dryness and delicious flavor is something really remarkable. Its outer color is a light blue. For price, see page 29.

I add extracts from letters received from various farmers and gardeners :—

"The Marblehead Squash seed I had of you produced a very fine crop of the first quality of fruit. I had 1200 pounds from 13 rods of ground, which was quite satisfactory."   J. M. MERRICH.
Wilbraham, Mass.

"The Marblehead Squash did the best of anything I ever tried to grow ; the neighbors all pronounced them first rate, and I think them better than any sweet potato I ever saw."   ROBERT STEWART.
Paulton, Westmoreland Co., Pa.

"The Marblehead is the best squash that I ever had. I raised sixty-three from the package—nice ones they were too. We cut the last on the 23d of March; it was so hard that I had to cut it with an axe. I recommend them to all."   S. D. GREENWOOD.
Wales, Me.

"THE 'MARBLEHEAD' SQUASH. Some weeks ago we received from Mr. James J. H. Gregory, of Marblehead, Mass., the introducer of the Hubbard, specimens of a new squash. The squash was tried by several, and unanimously pronounced to be of the very highest quality."
—*American Agriculturist.*

Henderson's Early Summer Cabbage.

This new Cabbage is much larger than the Early Wakefield, though not quite as early. This has a great and growing popularity among market gardeners, as a second early.

For a large extra early variety it is highly recommended. Per pkge. 15 cts.: per oz. 70; per lb. by mail, $7.00; by Exp. $6.85.

Paragon Tomato.

This fine new tomato *for four years in succession took the first prize in its class*, at the annual exhibitions of the Massachusetts Horticultural Society—where the critical standard is of the very highest character. *It ripens perfectly around the stem, and is the largest round tomato in cultivation.* The engraving making a good presentation of its characteristic shape. It is of good size and remarkably solid. In time of ripening it comes between the early and the late varieties.
Per pkge. 10; per oz. 50: per lb. by mail, $4.00.

Marblehead Champion Pole Bean.

After testing on my experimental grounds for the past ten years almost every variety of pole bean known, I find this new pole bean excels every other variety in *earliness.* While, as is well known to market gardeners, the pole varieties do not usually begin to blossom until the first picking has been made on the early sorts of bush beans that were planted at the same date, the Marblehead Champion Pole will be found to be so early as to have beans ready to pick or market *as early or earlier than the earliest bush varieties.* Per package, 15 cents.

Longfellow's Field Corn.

This fine field corn I have thought worthy to have its portrait taken. It is the result of careful selection in a family of Massachusetts farmers for forty-five years. The ears are remarkably long, some of them fifteen inches, and oftentimes two or more good specimens grow on one stock. The cob is quite small. It is the largest kerneled variety of yellow field corn that I have ever found it safe to plant in the latitude of Massachusetts. *Several of my customers have expressed themselves as highly pleased with their crops of the Longfellow Corn.* The seed I offer this season is selected from a crop of 220 bushels of ears to the acre. Per package, 10 cts.; per quart, by mail, 55 cts. by express, 35 cts.; per peck, $1.00 ; per bushel, $3.00.

Log-of-Wood Melon.

This new melon is a variety of the yellow-fleshed musk melon. It grows to the enormous length of *from two to three feet,* the form and general appearance of most of the specimens being very well shown in the above engraving, which was taken from a photograph. Early and prolific, in quality nothing extraordinary, but equal to the common Muskmelon. Per package 15 cts.

Vick's Early Watermelon.

I think so highly of this melon, particularly as an early variety, that I have had a specimen grown on my grounds, photographed and engraved. Of medium size, oblong, smooth, flesh bright-pink, resembling strongly the southern varieties, solid and sweet. I consider this one of the best of the early watermelons I am acquainted with.

Per package, 10 cts.; per oz., 20 ; per lb., by mail, $1.25 ; by express, $1.10.

Danvers Carrot.

In the town of Danvers, Mass., the raising of carrots on an extensive scale has for years been quite a business—the farmers finding a large market in the neighboring cities of Salem, Lynn and Boston. After years of experimenting they settled upon a variety which originated among them, (as did the Danvers Onion) known in their locality as the "Danvers Carrot." It is in form about midway between the Long Orange and Short Horn class, growing generally wit'ı a stump root. The great problem in carrot growing is to get the greatest bulk with the smallest length of root, and this is what the Danvers growers have attained in their carrot. Under their cultivation (see my treatise on Carrots and Mangolds) they raise from twenty to thirty tons to the acre, and at times even larger crops. This carrot is of a rich dark orange in color, very smooth and handsome, and from its length is easier to dig than the Long Orange. It is a first class carrot for any soil. The seed I offer is from carefully selected stock ; Per package, 6 cts ; per oz., 15 cts. ; per lb., by mail, $1 50 ; by express, $1.35.

Excelsior Melon.

This new melon is early, of large size and fine quality ; rind thin ; flesh, of a bright red color ; very delicate and sweet. Samples have been grown weighing over seventy pounds. It took the first premium at the annual exhibition of the Massachusetts Horticultural Society in 1877 and 1878, specimens being shown weighing sixty-five pounds. Per package, 10 cts.; per oz., 25 cts.; per lb., by mail, $1.50 ; by exp., $1.35.

[PLATE I.]

Early Schweinfurt Quintal Cabbage

Little Pixie Cabbage.

Early Ulm Savoy Cabbage.

Early Jersey Wakefield Cabbage.

Early Winn'gstadt Cabbage.

Premium Flat Dutch Cabbage.

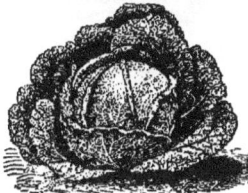

Improved American Savoy Cabbage.

Early Wyman Cabbage.

CABBAGES.

☞ For full particulars on Cabbage growing, see my Treatise, advertised in this Catalogue. ☜

For several years I have devoted two or three pages of my catalogue to quite a detailed presentation of my standard varieties of Cabbage and Squash. I do this because, having been the original introducer of several of these varieties, the public naturally look to me for the fullest explanation and description of them, and I therefore present these pages *for the information of the thousands of new customers who come for the first time each season, rather than for the perusal of old friends who from personal experience of years know all about their merits.*

MARBLEHEAD MAMMOTH.

MARBLEHEAD MAMMOTH. This is without doubt the largest variety of the Cabbage family in the world, being the result of extreme high culture. I have had heads, when stripped of all waste leaves, that could not be got into a two-bushel basket, having a diameter two inches greater! In a former circular I quoted from persons residing in fourteen States and Territories, and also in the Canadas, East and West, expressing their great satisfaction with the Stone-Mason and the Marblehead Mammoth Cabbages, in their great reliability for heading, the size, sweetness and tenderness of the heads. They had succeeded in growing the Mammoth to the weight of thirty and forty pounds, and in some instances *over fifty pounds!*

STONE-MASON CABBAGE. This Cabbage is distinguished for its reliability for heading, the size, hardness and quality of the heads. Under proper cultivation nearly every plant on an acre will make a marketable head. The heads vary in weight from nine to over twenty pounds, depending on the soil and cultivation. In earliness the Stone Mason is upward of a week ahead of the Premium Flat Dutch and makes a harder head.

STONE MASON.

FOTTLER'S
Improved Early Brunswick.

After an extensive trial on a large scale by market farmers in all parts of the United States, Fottler's Cabbage has grown in estimation, particularly in the great Cabbage districts of Long Island and in the vicinity of Albany, N. Y. My stock seed I have imported from the foreign seed growers from whom came the first stock sent to the United States.

Following will be found some of the recommendations I have received from those who have raised the Fottler's Cabbage. It is very rapidly growing in favor.

"From the seed of Fottler's Early Drumhead cabbage we raised cabbages that weighed 35 to 40 pounds. The rest of the seed did as it was recommended."   JACOB F. SELDOMRIDGE.
Ephrata, Penn.

"Your Cabbage seed have always proved most excellent. The Fottler's Early Drumhead is the best cabbage for general crop that I know of for this climate. With ordinary cultivation it is sure to make fine large heads. Last season I planted my Fottler's in the open air, in May, and raised very fine cabbage weighing from 10 to 20 pounds."
Pembina, Dakota Territory.   WM. K. GOODFELLOW.

"Those Fottlers were splendid. Every plant made a respectable head weighing from 8 to 25 lbs. each. I shall depend on you for what few garden seeds I use in the future."   S. J. WESTON.
West Peterboro, N. H.

"My cabbages were a perfect wonder to all who saw them. The great inquiry was where did you get the seed and what kind are they? Fottler's was my reply. From 15 cents worth of seed I have sold $50 worth, and have two or three hundred heads left yet. Cabbages were a general failure in this town except mine."
Westford, Mass.   E. J. WHITNEY.

FOTTLER'S IMPROVED EARLY BRUNSWICK.

"The Fottler cabbage is my favorite. It headed up uniformly and splendidly."   N. A. TAYLOR.
Houston, Texas, Feb. 9, 1874.

LITTLE PIXIE, EARLY ULM SAVOY, SCHWEINFURT QUINTAL. (For engravings see the previous page.) I recommend these three sorts as the best early Cabbages for family use. The first two are the earliest Cabbages grown, being each of them earlier than Early York. Little Pixie heads very hard, and all cook very tender and sweet; is earlier than Early York, and in many localities makes a first class market cabbage. The Savoys are the tenderest and richest flavored of all Cabbages, and for boiling are decidedly the best for family use, being much superior to the Drumhead and Cone-shaped varieties. Schweinfurt Quintal is decidedly the earliest of all the larger Drumheads; the heads attain to a diameter of from 16 to 18 inches, are very symmetrically formed, and are remarkably tender. When cooked they are very sweet, and quite free from any strong cabbage taste. They are so very tender they will not bear transportation in bulk any distance without serious injury; hence I recommend it as a capital Cabbage for early use in the family, but not as a market Cabbage, fitted for all localities, though very valuable for this purpose where the market is near at hand. (See engravings, Plate 1.)

IMPROVED AMERICAN SAVOY. This is probably the best of all the Savoys for the general market. It grows to a large size, is as reliable for heading as the Stone-Mason or Premium Flat Dutch, and has as short a stump as either of these varieties. I heartily recommend it to all those Market Gardeners who grow Savoys by the acre for the general market. (See engraving, Plate 1.)

EARLY WINNIGSTADT CABBAGE. No variety of early Cabbage, in my experience as a seedsman, has had a more regular and rapid growth in popularity than the Winnigstadt, which I attribute mostly to the fact that it is so remarkably reliable for heading even under very adverse circumstances. The Winnigstadt is also a large sized cabbage among the early kinds, and probably the hardest heading of all the conical varieties. In earliness it comes in about a week later than Early Oxheart. Should the soil of any of my farmer friends be of so sandy a nature that they find it extremely difficult to perfect any variety of cabbage, before bidding a final farewell to the cabbage family I would advise them to try the Winnigstadt. Planted in the latitude of Boston July 1st, the Winnigstadt makes a good cabbage for winter use. (See engraving, Plate 1.)

CANNON BALL CABBAGE. This Cabbage is so called because the head is as round, and almost as hard and heavy, as a cannon ball. I pronounce it as forming the *roundest*, *hardest* and *heaviest* head in proportion to its size, of any cabbage known. It matures about ten days later than the Early York. While about all varieties of early cabbage make rather soft heads, this, though early, makes the hardest heading cabbage known. The heads when fully grown attain to the size of from six to eight inches in diameter.

I present below a few extracts from the many letters sent from customers, relative to my Marblehead Mammoth, Stone-Mason, Cannon Ball, Winnigstadt, Schweinfurt Quintal, and Early Wyman Cabbages, etc.

"I send the weight of a part of the vegetables raised on this farm in 1877 from your seeds :—Red Drumhead Cabbage, 50 lbs.; Fottler's Drumhead Cabbage, 40 lbs.; Marblehead Mammoth Drumhead Cabbage, 50 lbs."—GEO. N. ENGLISH, *Sheridan, Clare Co., Mich.*

"Having tested your Early Wyman Cabbage two seasons, I wish to let others know that they are superior to any other variety I have ever grown. I have heads of this season's growth weighing 10 and 12 lbs. each, and the quality is excellent. They are the tenderest cabbage known in this town."—L. P. WALKER, *Union, Maine.*

"The package of Marblehead Mammoth Cabbage seed you sent me last spring did finely. I raised the largest heads of cabbage from them I ever saw grown in this country. They excited my neighbors and some are old cabbage growers."   J. W. CLOUSE.
Card Hill Post Office, Tenn.

"The Marblehead Dutch Cabbage that I had of you last spring was the best cabbage I ever grew; it appeared to be perfectly pure and headed up nearly to a plant."   W. L. CONOVER.
La Foyetta, Ind.

"Your seeds are splendid. We had splendid cabbage last year. Those Cannon Balls were the best cabbage that we ever had."
Unity, N. H.   BENJ. P. MARSHALL.

"We find your cabbages to be as good as you represented them to be. The Cannon Ball, Pixie, Stone Mason and Winnigstadt did splendidly. We had no success in raising cabbage until we began having seed from you."   Mrs. G. A. MORRILL.
West Alton, N. H.

"I must acknowledge that your early Wyman and Wakefield cabbages are the right kind for this hard woodland. This is the second year I have raised them. I find the one half was not told me, for I believe every seed came forth, and all are matured. I have already sold all I have. People all say I have the best lot of cabbage ever seen in this vicinity. They are also of good flavor and size."
Weare, N. H.   PAGE R. MERRILL.

"As this is, I think, the 8th season we have ordered seeds from your house, I feel it my duty to tell you that in no case were we disappointed in either name, quality or purity. All vegetables, and most especially Cabbages,—Mammoths, Stone Masons, Fottler's, Winnigstadt, and Cannon Balls presented a striking contrast to those generally raised in our neighborhood. Last year I ordered heavily and divided among friends who were astonished at results, and are now ordering of you for themselves."
Millersburg, Ind.   DANIEL LUTZ.

"My cabbages produced from seed purchased from you are the admiration of the whole neighborhood. In fact, all your seed are far superior to any ever introduced into this country, and any one has only to become acquainted with them to use no other"   J. M. FORD.
Spring Cottage, Miss.

"I raised the past season, from seed purchased from you, heads of Fottler's Early Drumhead that weighed 45 pounds apiece. I had an acre of Fottler's and Premium Flat Dutch that were the best lot I ever saw together. Scarcely a plant failed to head, notwithstanding the green worms were very bad about here."   JOHN D. MILLER.
Elmira, N. Y.

"The seed you sent me last year gave great satisfaction, particularly your Marblehead Cabbage. They grew to the weight of 44 lbs. and 2 oz."   SAMUEL BAKER.
Ottawa, Canada.

"You sent me a package last year of the Early Schweinfurt Quintal, and it was the best cabbage we ever raised. I could sell it readily at twenty-five cents per head. Some heads weighed 36 lbs."
Orbisonia P. O., Pa.   GEORGE SWARTZ.

"I have got out and heading up nearly 40,000 cabbages from seeds obtained of you—the largest cabbage crop ever grown in the state. They are looking splendidly."   H. M. STRINGFELLOW.
Galveston, Texas.

"Your Wyman Cabbage is *the* Cabbage for this place. I transplanted in April 300 plants, commenced to sell on 12th of July; on the 25th of September had sold 2550 lbs. at 5 cts. per lb. They weighed from 5 to 13 lbs. each. I do not write this as an advertisement, but to say that I have faith in your seeds."   CHARLES MAYNARD.
Bay Fork, Cal.

☞ My customers at the *South* will please observe the following : ☜

"Your Stone Mason cannot be beat for early Spring. I have raised them to weigh 16 lbs. Our inhabitants never saw such large ones grow South before."   J. S. STEBBINS.
Bloeboro, Ga.

"Mr. H. J. Van Pelt, of Mandarin Point, Florida, has been very successful in vegetable raising. Yesterday he deposited in our office a cabbage which weighs 38 pounds. It is solid and perfectly formed, of the Marblehead Mammoth variety. The seed were sown in September last, transplanted in October on an area of three-fourths of an acre, fertilized with 500 pounds of Fish Guano, procured of Mr. J. W. Hawkins, of this city, composted with swamp muck, and applied broadcast and in the hill. He commenced marketing the first of April and finished yesterday. The cabbage have varied from 10 to 20 pounds in weight. Total receipts from three-fourths of an acre, over $400. The seed was procured from Mr. Gregory, of Marblehead, Mass., who makes cabbage a specialty."—*Jacksonville [Florida] paper.*

"I think it would be a difficult matter to find a finer lot of Cabbages than those I have growing from the Little Pixie, Cannon Ball and Winnigstadt seeds obtained from you in the winter."
Walterboro, S. C.   W. S. HARLEY.

"The paper of your Mammoth Cabbage seed sent me last summer was duly received, and from them I raised the largest and finest cabbages that I have ever grown, in a trial of about 40 years."
Griffin, Ga.   J. S. JONES.

Cream-Fleshed, Sculptured-Seeded Melon.

In size about medium, color much like Phinney's, but darker and more regularly striped; flesh very tender and melting, sweet and delicious. Melons quite thin-shelled, but first-rate keepers notwithstanding. The seed present a singular and striking appearance, as though engraved with oriental characters. Price 15 cents per ounce; 6 cents per package.

Very Large White Russian Winter Radish.

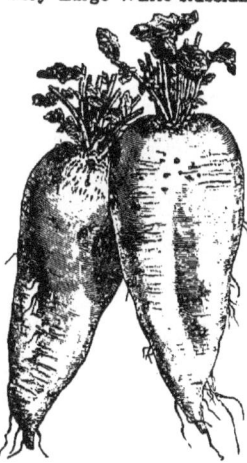

The largest of all the winter sorts. From seed sown in June (for winter use the latter part of July is better) roots can easily be raised to weigh three pounds each. To obtain the best results the soil should be made rich, light and pliable. In the absence of rain, water freely. For winter use pack the roots in earth or sand out of danger from frost. Immerse for a short time in cold water before using. To be used as a salad or served in all the ways of the spring and summer radishes. For prices see page 28.

The Hubbard and American Turban Squashes, &c.

THE HUBBARD SQUASH. As the original introducer of the Hubbard Squash, I offer to the public seed taken from squashes

THE HUBBARD SQUASH.

grown specially for seed. Let me not be understood as saying my seed is *perfectly* pure, for, as every farmer who has had experience knows, *perfect* purity in the Squash family is impossible ; but the seed I raise is nearly perfectly pure,—as near as the utmost care by isolation for years can make it, and the great difference in purity between this and the average seed of the market, every farmer who has tried it knows.

AMERICAN TURBAN SQUASH. I have sent this fine Squash out as the best of all fall squashes, as good for fall as the Hubbard is for winter.

I have full faith that the Turban will soon be adopted throughout the United States as the best of all *fall* Squashes.

☞ I note that by one or more of the Philadelphia seed firms the Turban Squash is spoken of as a showy variety of but little value for domestic use. It is very evident that they have confounded the showy but worthless *French Turban* with the *American* Turban Squash. I give extracts from a few letters which show how the American Turban is appreciated.

"I think your Turban Squash is the best fall squash I ever ate. We had them till January, and my neighbors that tried them said they were the best squashes they ever saw." LEVI B. SIBLEY.
Windsor, Me.

"I unhesitatingly pronounce the Turban Squash the very best squash that grows. I have kept them into January." MRS. MARTHA WOLF.
Greenvale, Iowa.

"I have had Turbans this month (March), just as good as they were last November."—A. W. VALENTINE. *Bethel, Maine.*

"The American Turban Squash is just the thing it is represented to be. One squash is worth a dozen eggs in the way of making egg custards, and no person should be without them that likes something good to eat." S. E. RANKIN.
Shady Grove, Washington Parish, La.

MAMMOTH SQUASH AND SQUASHES FOR FEEDING TO STOCK. Mammoth Squashes, though of but little value for table use, on rich land in those sections where roots are but little cultivated, are very profitable as food for cattle. I give the substance of letters received from a few of my correspondents, that my friends may be able to compare notes. A half acre of these Squashes have averaged 75 lbs. each in weight. Among prolific varieties for stock, the Vegetable Marrow holds high rank. I have raised fourteen tons on a single acre of land.

"The seeds received from you last spring gave good satisfaction. Among other seeds received from you was one package of the Mammoth Yellow Chili Squash, which grew to the weight of 192 lbs. Was of a handsome shape and of a deep rich orange color." JAMES B. PICKERING.
Portsmouth, N. H.

"I got two seeds of your Mammoth Chili Squash, and from them I raised nine squashes weighing 640 lbs. The largest of them weighed 236 lbs." H. Y. DREMER.
David City, Nebraska.

"One vine of Mammoth squash from seed of you gave us one squash weighing 164 lbs., another of 145 lbs., and several of 30 to 50 lbs." J. O. NOTESTEIN.
Canton, Ohio.

"From the package of Mammoth Yellow Chili Squash seed had of you last spring, I can report several squashes weighing over 100 pounds each, and one weighing two hundred and thirty-eight pounds (238), this, too, notwithstanding a very cold, wet season." W. HESTER.
Vanbreter, Iowa.

"I had good success in raising Mammoth squashes last year from seed I got from you. I had three that weighed 173, 137 and 129 lbs., and several others from 75 up to 104, all from three vines." JAMES ARNOLD.
Farm Hill, Minn.

Canada Victor Tomato.

This remarkably early tomato which I introduced six years ago will be found to excel in the most desirable characteristic ☞ *in earliness of ripening the great bulk of the crop,* ☜ a trait of great value to the market gardener.

The fruit is in nearly all locations of large size, generally symmetrical and handsome, while in ripening it has no green left around the stem, a great fault with many kinds otherwise good. The fruit is heavy, full meated and rich, between round and oval in shape, and red in color; it is distributed very evenly on the vines.

I offer headquarters seed this season by the package, ounce

and pound. For seed of my own growing, saved from selected specimens—per package, 10 cents : per ounce, 45 cts. ; per lb. $4.50. I have again had a special selection of seed stock made for me, from a crop grown in Canada, by the originator ; a few bushels of the very earliest being selected for seed from a field of some acres. This seed is 15 cents per package, and $1.25 per ounce. Dealers supplied at a discount.

What the public have very generally found it to be, will be pretty conclusively shown from the testimonials that follow for which I have to thank my customers,—all free-will offerings. I regret that want of room will allow me to publish but few of them.

"The seeds I had of you last year all proved good. I had ripe Canada Victors the 10th day of June." W. H. BYER.
North Reading, Mass.

"I planted seven varieties in my hot bed at the same time. All had the same treatment, and were transplanted to the open ground on the same day. All were situated exactly alike in the field, having the same treatment. I picked the first ripe Tomato on the 12th of July, from the Canada Victors. From that time they began to ripen regularly, and on the 20th of July I picked 16 ripe tomatoes of good size from one vine. On the 17th of July I picked the first Trophy, and also the first Hathaway's Excelsior; July 23d, Gen. Grant and Essex Early; July 25th, Burton Market, and July 26th, Arlington. I had about 350 Victor plants, and about 2000 of all other varieties. From the time of picking the first tomatoes to the 1st of August, I had picked more ripe tomatoes from the Canada Victor 300 plants, than from the 2000 of the six other varieties. The Victor thus proved to be beyond all comparison, the most profitable early tomato." R. W. HARGADINE.
Felton, Delaware.

"The Canada Victor is fully a week ahead of anything I have tried, and I have tried nearly all kinds. The Early York and Hubbards Curled are the earliest with me, but are fully a week behind Victor. I have tried Orange Field, Maupay's Superior, Cedar Hill, Keyes' Prolific, Cook's Favorite, Hathaway's Excelsior, and, in fact, every kind that I thought was likely to succeed. The Victor Tomato is a very solid tomato, weighing 60 lbs. to the bushel, plump weight. It is first class in point of flavor. Moreover it has the remarkable quality of keeping its flavor late in the season; what I mean is, that when ripened in cool, wet weather, it is of fair flavor when other varieties are quite insipid, and almost or quite useless. This, in part, is accounted for from the fact that it seldom cracks open like other tomatoes, but keeps sound. This is a very valuable feature to us here, where our seasons are short, and are troubled with frosty nights, sometimes every month in the summer. If you wish to make use of any of my statements you are at perfect liberty to do so." S. H. MITCHELL. *Ontario.*

"The Conqueror with us has proved early, smooth, and very productive, but in size and solidity is inferior to the Canada Victor, which latter, all things considered, is yet our *best early* tomato for market or home use."—TILLINGHAST BROS., *La Plume, Pa.*

"I tested your Canada Victor Tomato grown from your extra selected seed, with Conqueror and Keyes' Prolific for the first time this year. It was fully a week ahead of Conqueror, and at least 10 days ahead of Prolific.—E. W. HARGADINE, *Felton, Kent Co., Del.*

"I must say a few words about your Canada Victor Tomato. I bought of you two papers and realized over fifty dollars on the two papers. They are two weeks earlier than any I have ever planted." *Kennaysville, West Va.* WM. SHAGRUDEN.

"CANADA VICTOR. We tried this new tomato last season, and our opinion is that it will occupy the same place as an early variety that the Trophy does as a medium and late sort. It is the most uniformly smooth of any of the flat varieties—is very solid and ripens up to the stem without a green spot. We expect to have a good many acres in tomatoes the coming year, and shall plant but two sorts, Canada Victor and Trophy. We are satisfied that we have found the best sorts for our use, and shall experiment no more."—*W. F. Massey in Am. Farmer.*

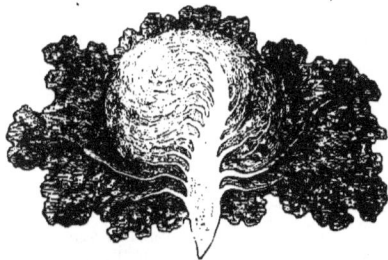

✗ THE HANSON LETTUCE.

The above cut represents a sectional view, showing the inside of this truly superior Lettuce, the heads of which grow to a remarkable size and are deliciously sweet, tender and crisp, even to the outer leaves. A single head is frequently large enough for an ordinary sized family. The color is of a beautiful green without, and white within. This variety is free from any bitter or unpleasant taste found in some sorts. The heads weigh from 2½ to 3 lbs., and measure to outer leaves 18 to 25 inches in diameter. It is not recommended for forcing, but for outdoor cultivation it cannot be excelled, if grown in well manured and cultivated ground. Per package, 15 cents.

"Your seeds give satisfaction, especially the Hanson Lettuce. I had some heads that weighed 3 lbs., trimmed." CARL A. BESCH. *Monee, Will Co., Ill.*

"Myself and neighbors agree that the Hanson Lettuce is the finest we ever saw. One head is plenty for a large family at one meal." *Wesley, Tex., May 18, 1874.* GREGOR C. McLEOD.

"As to the Hanson Lettuce, you don't half praise it. I raised one head that measured twenty-five and a half inches in diameter. *Bloomingdale, N. Y., March, 1875.* MRS. LOUISE M. LENNON.

Tailby's Hybrid Cucumber.

Mr. Tailby made this choice new variety by his skill and perseverance in crossing the Early White Spine on one of the largest of the English Frame varieties. These English Frame cucumbers are much larger than our American varieties, some of them growing to thirty inches in length, but they are so tender that they cannot usually be relied upon in open air cultivation in this country. In Tailby's Hybrid however, we have a perfect success in hardiness, as it proves to be equally hardy with our American varieties. In size it is larger than White Spine, while it retains all the smooth beauty of its English parentage. It is literally an enormous cropper, and for size, beauty and number it is a sight to behold. Price, 30 cts. per ounce, 10 cents per package.

The editor of the *Rural New Yorker* who has been testing this with other varieties on his extensive experimental grounds gives the following as the result:

"The results of our cucumber experiments this season, may be summed up in a very few words. We choose Tailby's Hybrid for cucumbers and Green Prolific for pickles."

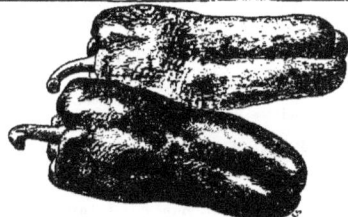

SPANISH MONSTROUS PEPPER.
(See page 28.)

✗ HARDY RIDGE, OR PRESCOTT MELON.

Probably not one person in a hundred seeing the Hardy Ridge when growing would take it for a melon. Nevertheless it is a melon, and one of the *very best quality*, too. A very popular variety in the markets of London and Paris, the wonder is it has not before this been introduced into the United States. It is prolific, grows to a very large size, is of splendid quality,—being equal to the best cantaloupes, while it is *by far the thickest meated of all melons*, being in fact, very nearly solid and having but very few seed. Price, 15 cents per package; per oz., 40 cts.

RARE, NOVEL. OR VERY DESIRABLE.

Among the New, Rare or Desirable Vegetables, I would call particular attention to the following.

☞ While most of these new and rare vegetables will be found to be of universal value, some may vary in quality with the soil and locality, and the value of others vary with the varying tastes of my customers. As a general rule we are not rendered capable of passing judgment on a new vegetable by the result of a single trial. Oftentimes the most we learn from the result of planting one season is what are *possibly* the merits or demerits of it; a second may develop what are *probably* its merits or demerits; and usually a third season will be required to enable us fully to determine its value, and give the new-comer its true place in the vegetable garden. Take Mexican Sweet Corn for an example; should the first season of experiment be wet and cold at the time it matures for table use, this variety being more sensitive in its habits, than the old standard sorts, may be more affected in its quality than they, and not superior to them in sweetness. Now let the next season be a hot and dry one, and the same corn, having a season more congenial to its tropical origin, will develop its full quality and demonstrate its full claim to the rank given it in my catalogue. So with many of the varieties of our Tomatoes; from an extended cultivation of many kinds, I am fully convinced that though some have been overpraised, yet with a majority an experience of three years would reverse or greatly quality the hasty opinious often expressed of them, from a trial of but a single season.

☞ **New Vegetables for 1880.** The following are the more rare and valuable vegetables which I introduce into my catalogue for the first time this season.

☞ *For prices per bushel, pound, quart, &c., please see pages 19–31 inclusive.* ☜

DEFIANCE ASPARAGUS. Smalley's Extra Early Defiance Asparagus grown side by side with Conover's Colossal and subject in all respects to the same treatment, grew full twice as large besides being a week or two earlier. It is of a rich green color and excels every other variety thus far known in tenderness, and has no superior in flavor. It will be large enough to cut a year sooner than other varieties. Two-year old roots per hundred, $5.00 .... | 25

HULLESS BARLEY. (See third page of cover.) | 15

CRYSTAL WHITE WAX BEAN. A variety of wax bean, prolific, but in my experimental test grew too near the ground to make it desirable. The York Wax and Dwarf Black Wax are better sorts in this respect .... | 15

GOLDEN BUTTER BEAN. A new, French, wax pole, early and prolific. This bean closely resembles the Indian Chief, but it is a better bearer and the pods are rather longer .... | 15

YELLOW PODDED WHITE WAX BEAN. This is a very superior early pole wax or stringless bean. The pods are as long as Giant Wax, but it *surpasses this old variety in earliness and productiveness* .... | 15

DWARF RUSSIAN BEAN. A new bush bean sent out by Messrs. Vilmorin & Co., of France. It is a singular looking bean, but thus far I have found nothing to note wherein it is superior to our standard sorts .... | 10

CHINESE BEAN. (See page 2.) .... | 13

ROSE BEAN. (See page 3.) .... | 13

ECLIPSE TURNIP BEET. I imported last season all the varieties of beets grown in Germany, and the Eclipse gives me more satisfaction than any of the others. I think it will prove an acquisition to our American gardener. The top is small; the beet itself is in form much like Bastian's Early, while the flesh is of a good dark color. One of the early sorts .... | 10

GERRY ISLAND CAULIFLOWER, New. (See page 2.) | 30

LATE ALGERIAN CAULIFLOWER. One of the new, large, French varieties. Very fine .... | 20

BERLIN DWARF CAULIFLOWER. From a test made side by side with many other varieties the past season, I find that the Berlin Dwarf, for earliness, size and quality, ranks about with the Early Snowball .... | 25

CALIFORNIA OR GOLDEN BROOM CORN. This is superior to the Evergreen for many purposes. It grows a very long brush which never gets crooked, and yields more tons of brush than any other sort .... | 15

BLUNT'S PROLIFIC CORN. (See page 3.) .... | 10

EARLY BOYNTON SWEET CORN. This ranks very high in the west as a valuable early sweet. Our western

friends will find Marblehead Early decidedly ahead of this or any variety I have yet found in earliness ...... | 10

SHORT FRENCH PICKLING CUCUMBER. A new, French variety, closely allied to Green Prolific, making an excellent pickling cucumber .... | 10

NELLIS' PERPETUAL LETTUCE. A distinct variety being exceedingly tender and rich It remains fit to cut up to seeding, while it takes the longest time before running to seed of any sort, It does not head, but makes a huge, compact, bushy growth. .... | 13

CABBAGE LETTUCE, (*Pellacier*). Heads very large, spherical, very solid, and of a fine, tender, taste. Leaves deeply cut and irregularly notched, giving the plant a very decorative appearance May be used for forcing or for out-door cultivation .... | 20

CHICAGO NUTMEG MELON. This variety grows to a larger size than the Boston Nutmeg, and hence is very popular with those who lay great stress on size for market purposes .... | 15

PERSIAN MUSKMELON. One of the deliciouly flavored, salmon-fleshed class. It is rather later than Ward's Nectar, grows rather larger and is very thick fleshed .... | 10

ESSEX HYBRID SQUASH. This is a cross between the Turban and the Hubbard, having the shape of the former and the shell of the latter. It appears to combine the qualities of the two parents. The characteristics are said to be permanent and the squash itself to be earlier than either the Hubbard or Turban. Should this be so it will be a decided acquisition to our list of squashes. Not having as yet cultivated this variety, I am not able to speak from experience. Per oz. 40 cts; per lb., $8.00 .... | 15

TURK'S TURBAN TOMATO. (See engraving, page 3.) | 20

CHAMPLAIN WHEAT. A Spring wheat made by a scientific crossing of the Black Sea and Golden Drop, having the beard of the former and the white chaff of the latter, free from rust and smut, and giving a flavor of superior quality. The straw is strong and vigorous, standing erect and frequently bearing heads 5 to 6 inches in length, containing 60 to 75 kernels each. Price per pound by mail, 50 cents; 3 lbs., $1.00; by express or freight at purchaser's expense; per peck, $3.00; per bushel, $10.00. .... | 10

DEFIANCE WHEAT. Another of Mr. Pringle's new hybrids. It is a beardless white chaff wheat with long heads closely set, with large, white kernels, frequently numbering 75 to 80 to the single head. Early and characterized by great vigor. The straw is stiff, white and erect. Price per lb. by mail, 50 cents; 3 lbs., $1.00; by express or freight at purchaser's expense; per peck, $3.00; per bushel. $10.00. .... | 10

	Price per P'k'ge

EXTRA LONG SMOOTH CUCUMBER. A long, smooth, very straight, frame variety. The frame varieties grow much larger than our common garden sorts..   15

WHITE GERMAN CUCUMBER. (See page 3.).......   25

LONG GREEN SMOOTH CUCUMBER from Athens. This new sort is sent out by a celebrated German seed firm. It is one of the remarkably long frame varieties, and what is rare with all that class does well in the open air in this country   15

DANDELION. Very double. A new French strain sent out by Messrs. Vilmorin & Co., and especially recommended to market gardeners............   15

THE SURPRISE MUSKMELON. This new melon has a thin, cream colored skin and a thick, salmon colored flesh. Early, very productive, and of delicious flavor. Externally it resembles White Japan, *but grows to twice the size.* A first class melon..........................   15

BAY VIEW MELON. New. (See page 2.)...........   15

WHITE ZEALAND OATS. (See 3d page of cover)...   10

DR. McLEAN'S PEA. Our English friends declare this to be a splendid pea, even superior to Advancer. A wrinkled variety, large podded, with 8 or 10 peas to a pod. If it proves to surpass the Advancer it will be a great acquisition   15

ALPHA TOMATO. New. (See page 3.)...........   20

TRIUMPH TOMATO. A large sized, productive, solid sort, ripening well around the stem. With me it grew rather rough this season, but it is said to be generally a very smooth variety....   10

RED CHIEF TOMATO. A new variety. A cross between General Grant and Excelsior; thrifty and productive; foliage large and thick; fruit of good size and regular in shape; solid and with but few seed; rather late..   15

NEW JAPANESE TOMATO. A new variety sent out by an eminent German seed firm. My crop this season developed nothing in the earliness or appearance of it that was especially desirable......................   10

CHINESE YAM (*Dioscorea Batatas*). Allied to the potato, but containing more starch; quite hardy, remaining in ground over winter without protection. May be boiled or roasted, being quite farinaceous, nutritious and valuable for food. A very rapid grower, making a fine, ornamental vine, sometimes called cinnamon vine from the peculiar odor of the blossoms. The tubers increase in size from year to year. Packets of small bulbs..   20

The following are the more rare or choice of the select varieties of vegetables of former years, with prices per package. Prices per bushel, pound, quart, &c., will be found on pages 19 to 31, inclusive.

ALFALFA OR LUCERNE. This has until recently been considered too tender to stand our northern winters. Mr. Albert Chapman and Solomon Jewett have each met with fine success with it in Vermont, and Mr. C. cut four heavy crops from it in a single season. The success appears to turn on using American grown seed and planting it in a deep porous soil. Doubtless a top dressing with fine manure would serve to help it through its most tender period the first winter........   10

DWARF GOLDEN WAX, *alias* YORK DWARF WAX BEAN. In the year 1871 I introduced this bean to the public, giving it the name of "York Dwarf Wax." It proves to be more prolific, and yields larger and broader pods than the common kinds of dwarf wax beans. It is entirely stringless, threshes out easily, and is also a good shell bean for winter use....   10

YARD LONG BEAN. A curious bean of very dark and glossy foliage. The pods grow two-feet and upwards in length........   15

KENTUCKY WONDER BEAN. The most productive variety, that I have ever known. The vines take to the pole exceedingly well and the pods grow in clusters of three or four, being remarkably long, round and pulpy, covering the poles from top to bottom. I do not recommend it as a shell bean, but as a snap bean it is a "Wonder" as every market gardener will find. The pods are nearly a foot long, yielding from eight to ten beans to each pod..   15

LAMBERSON'S WHITE BEAN. Mr. Lamberson's new bean is remarkably prolific; pods of good length and snap well. The beans are *white* in color. A capital sort for marketmen who want a bean that will give an immense crop of good snap beans.   15

BASTIAN'S EARLY BLOOD TURNIP BEET. A new Beet, earlier than Bassano with a beautiful blood-red color when boiled. Very handsome in shape. This new beet has grown rapidly in popularity with market gardeners, and is generally preferred to the Red Egyptian.......................................   6

CRANE'S EARLY WYMAN CABBAGE. Market gardeners will find this very solid strain of the Wyman Cabbage an improvement in making harder heads than the common sort.........   15

MARBLEHEAD MAMMOTH CABBAGE. For a description please see page 10..   10

MARBLEHEAD DUTCH CABBAGE. My customers will find this an improvement on the common Flat Dutch, in its producing a remarkably symmetrical, round and handsome head; firmer and harder than the ordinary Flat Dutch..............   10

VILMORIN'S EARLY FLAT DUTCH CABBAGE. This is the French strain of the Early Flat Dutch, the heads being rounder and harder than in the strains of this early cabbage as grown in the United States...................................   10

EARLY BLOOD RED ERFURT CABBAGE. A new sort from Germany. Heads of an intense blood red color, very hard, and weighs from 12 to 14 pounds. The leaves are smaller and spread less than those of the common varieties of red cabbage.......   25

HENDERSON'S EARLY SUMMER CABBAGE. See page 7.....   15

HEARTWELL EARLY MARROW CABBAGE. A distinct and excellent variety of an early Cabbage, both for the garden and for marketing purposes. The heads are extremely firm, weighing from 4 to 6 lbs., with scarcely any loose outside leaves, the flavor particularly mild and melting.........................   15

ALGIERS CANTELOPE. Allied to the Hardy Ridge—but longer in shape. Flesh remarkably thick, the melons being nearly solid. In quality superior to the yellow fleshed muskmelons..   20

EXTRA EARLY DWARF ERFURT CAULIFLOWER. (Seed specially selected.) Very early, hardy, dwarf and compact; larger than Walcheren. The best for forcing and for general purposes as an early variety. Its compact habits admit of a large number being raised on a given area. I grew some heads for seed stock this season, that measured 18 inches in diameter.   50

EARLY SNOWBALL CAULIFLOWER. Very early and very reliable for heading, besides being very dwarf in its habits of growth, and with short outer leaves, thus allowing planting 20 inches apart each way...................................   50

CRAWFORD'S CELERY. (See page 4.).................   10

BOSTON MARKET CELERY. This is the short, bushy, compact, solid celery, for which Boston Market is so famous....   10

CHUFAS. These very closely resemble in sweetness and richness of flavor a cocoa nut. Very prolific, a single one yielding from two to four hundred. Plant one foot by eighteen inches   6

WATER CHESTNUT. To be grown in muddy brooks, the form of the Water Chestnut is very beautiful, and as it can be kept indefinitely it makes an elegant little ornament for the parlor table   10

PRICKLY COMFREY.. (See page 6.)................

BISMARCK CUCUMBER. A cross between the White Spine and Long Green, of even color, straight, tender, crisp and of fine flavor..   15

GREEN PROLIFIC PICKLING CUCUMBER. Selected with great care by one of the largest growers of pickling cucumbers in the country...   10

	Price per P'k'ge
MARBLEHEAD MAMMOTH SWEET CORN. My Mammoth has taken the first prize at two of the Annual Exhibitions of the Massachusetts Horticultural Society, the ears exhibited weighing, as gathered from the stalk, between two and three pounds each. This is a very sweet corn for family use. It is the earliest, sweetest and largest of all the Mammoth Sweet varieties. I offer packages from selected ears............................	15
PRATT'S EARLY CORN. One of the earliest of marketable size. It is an acquisition for marketmen as an early variety to come in before Crosby's, or Moore's. The ears are of a fair market size, well filled............................	10
LONGFELLOW'S CORN. (See page 8.)............................	10
MARBLEHEAD EARLY SWEET CORN. (See page 4.)............	15
RUSSIAN NETTED CUCUMBER. A native of the Ukraine country and very prolific; surpasses all others in hardiness; middle size, flesh white, and the skin covered with a pretty brown network which imparts to the fruit a peculiar appearance............................	10
CHINESE LONG NETTED CUCUMBER. It is a fine long variety with attenuated neck and prettily reticulated skin. Very prolific and hardy. Flesh thick and firm............................	25
TAILBY'S HYBRID CUCUMBER. See page 13............	10
NOURITON GIANT CUCUMBER. This is one of the frame varieties, sent out by the English seedsman as "The finest, longest and most prolific cucumber cultivated." I have grown them longer than an ordinary flour barrel............................	20
SNAKE CUCUMBER. A very long variety, (I have raised them six feet in length,) growing coiled up, having much the appearance of a large snake with the head protruding. Fine specimens will sometimes being $3.00 each as curiosities............	20
MARQUIS OF LORNE CUCUMBER. Messrs. Carter & Co. describe this new frame cucumber as follows :—"Of great value for exhibition purposes. It has a beautiful short neck, smooth skin, is very straight and prolific. It has gained many valuable awards."............................	50
VERY EARLY DWARF EGG PLANT. A new French variety of Long Purple. Earliness in the egg plant family is exceedingly desirable, and in this new sort we have an acquisition....	15
BONNET GOURD, DISH CLOTH GOURD OR LUFFA. The peculiar lining of the fruit, so tough, elastic and enduring, has given it its name. The vine is very ornamental, having dark green foliage with silvery shade, and large yellow blossoms in clusters. It requires a frame or support. The seeds should be started in hot bed. As the name indicates it is sometimes used as a dish cloth............................	15
ORNAMENTAL GOURDS. The packages contain seeds of Apple, Orange, Pear, Quince, Bottle, Egg, and other varieties. Peculiar, attractive and ornamental. Don't manure too high.	10
SUGAR TROUGH GOURDS. These grow to the capacity of several gallons, and will last years as sap vessels, or for holding liquids............................	12
ALL-THE-YEAR-ROUND LETTUCE. Very hardy, crisp eating, and compact. May be sown for succession all the year round. Does not tend to seed............................	15
SATISFACTION LETTUCE. A new English variety, large and unusually tender, remaining in head a long time............	15
EGYPTIAN LETTUCE. A large, spreading summer variety; color light green tinged with brown; very handsome. Not inclined to run to seed. Quality first rate............................	15
STONE-HEAD GOLDEN YELLOW LETTUCE. A new variety from Germany. As early as the White Tennis Ball, with larger heads—quality, first class; the decision of several experienced gardeners was, that they had never eaten anything of the lettuce kind that surpassed it............................	15
LOG-OF-WOOD MELON. (See page 8.)............................	15
SILVER-NETTED MUSK MELON. Very productive, uniform in size and high flavored. Holds a high rank in the West. ...	15
HARDY RIDGE MELON. (See page 13.)............................	15
SILL'S HYBRID MUSKMELON. This has all the earliness and sweetness of the White Japan, but is more spicy and delicious. Very vigorous and productive. The flesh is of salmon color. No garden should be without it............................	6
EXCELSIOR MELON. (See page 8.)............................	10

	Price per P'k'ge
GOLDEN FLESHED WATERMELON. The flesh of this melon is of a rich honey color and the flavor sweet and rich. Slices alternating with those of the scarlet fleshed variety make a pleasing show on the table. Shape nearly round. Size above the average............................	15
PHINNEY'S WATERMELON. This is one of the best early varieties I have introduced. Red-fleshed, early and excellent. Those who could not succeed with other kinds of watermelons have succeeded well with this. It stands transportation remarkably well............................	6
SCULPTURED SEEDED CREAM FLESHED WATERMELON. A new melon from Japan, which I introduced a few years ago. The flesh, which is very sweet, is of a delicate cream color. The seed are singularly sculptured with marks resembling oriental characters............................	6
PEARL MILLET (*Penicillaria spicata*). This new forage plant will not give satisfaction unless planted on warm soil after the ground has become heated by the sun's rays. On warm land, highly manured, yielded per acre last season as follows : At first cutting 45 days after planting, when 7 feet high, 30 tons of green and 6¼ tons of dry forage. A second cutting 45 days later, when 9 feet high, 55 tons green and 8 tons dry forage. At third cutting, Oct. 1st, 10 tons green and 1¼ tons dry forage, the aggregate being 95 tons green or 16 tons dry fodder within 135 days. It ranks about with corn fodder, and is readily eaten when either green or dry, by horses and cattle. Two quarts are sufficient for an acre in drills, or four when sown broadcast....	10
CHINESE HULLESS OATS. These thresh directly from the straw, as clear as wheat, without a particle of hull or chaff adhering, the grain being much larger than common oats when hulled, and weighing as high as fifty-five pounds to the measured bushel. Earlier than the common varieties. Authorities differ as to their value when compared with the common oat.	10
NEW QUEEN ONION. I find that this new onion is decidedly the earliest of all varieties. It does not grow to a very large size, but sufficiently large to bunch while green, and with extra liberal manuring I doubt not it will make a good sized onion when dry. Grown from seed, it was two inches in diameter by the 15th of July............................	10
GIANT ROCCA ONION. This is a Mammoth Onion sent out by one of the English seed firms. To get the fullest development of size it should be grown for sets the first season, then stored to be planted for a second season's growth............	10
WHITE GLOBE ONION. This, when well grown and well cured, is the handsomest of all onions, bringing at times double the price of any other sort in the New York market. It requires a long season in the North, and to keep the color pure, white should be pulled as soon as it begins to dry down and be cured in the shade, say in a barn where there is a good draught of air............................	10
FERN LEAVED PARSLEY. A most beautiful thing. Valuable as a decorative plant. Resembles a beautiful moss......	10
HANCOCK EARLY PEA. A new American pea of the first early class. See notes elsewhere............................	10
LAXTON'S SUPERLATIVE PEA. Messrs. Carter, the English seedsmen, speak of this new pea as follows :—"The largest and finest podded pea yet raised ; indispensable as an exhibition pea ; pods have been grown 7 inches in length and in some larger than the parent pea, Laxton's Supreme, which has taken the first prize for several years." Second early, color and flavor unsurpassed............................	15
CULVERWELL'S TELEGRAPH PEA. Messrs. Carter & Co., the English seedsmen speak of this new Pea as follows :—"This is an extraordinary acquisition, the Peas often being so close together as to appear to be forming a double row in the pod. It is likely to be the forerunner of a new type of this indispensable summer vegetable."............................	20
GOLDEN YELLOW SUMMER TURNIP RADISH. Its shape is that of the Yellow Summer Turnip Radish, but the root is more spherical, its neck is finer and the leaves are smaller. Of very rapid growth, it is fit for use from 4 to 6 weeks after having been sown. A novelty of great merit............................	10
WHITE EGG TURNIP. (See page 5.)............................	10
TEOSINTE. (See page 5.)............................	15
WHITE RUSSIAN SPRING WHEAT. (See page 5.)............	10

CARTER'S CHALLENGER PEA. Messrs. Carter & Co., the distinguished seedsmen of England, highly recommend this new pea as being one of the handsomest, most prolific, and best flavored varieties in cultivation. It is a magnificent exhibition Pea, and will speedily find favor amongst growers for market by reason of its fine, handsome pods, productiveness and dwarf habit. It is a dwarf, dark-green marrow, growing about two to two and a half feet in length, and the entire haulm is literally covered with pods...................................... 20

NEW GOLDEN PEA. The pods of this new Pea are of a delicate yellow when sufficiently matured for green-shelling for the table—the Peas also being of a delicate straw color. Good cropper. Unique...................................... 15

CARTER'S LITTLE WONDER PEA. Of this new English Pea. Messrs. Carter & Co. state, "We are satisfied that a trial of this Pea will fully establish its title to be called a Little Wonder, whilst its remarkable qualities will speedily insure its universal cultivation. It is best described as a wrinkled marrow, as early as the Advancer, with pods like the finest type of Veitch's Perfection. Height 20 to 24 inches, very robust habit, wonderfully prolific and of fine flavor. We consider it distinct, desirable and likely to supersede the Advancer, the seed being larger, and the Pea quite distinct from it.".................. 20

CARTER'S COMMANDER-IN-CHIEF. A grand Pea for exhibition and general purposes of cultivation. It is a green, wrinkled marrow of exquisite flavor, with fine, slightly curved pods sometimes containing ten large peas. During the past two seasons the long and handsome pods of Commander-in-Chief, carrying a beautiful bloom have been prominent at the principal Vegetable competitions in England........................ 15

SUTTON'S EMERALD GEM PEA. This new first early pea is quite distinct from all others, and is undoubtedly one of the finest varieties in cultivation. The peas retain their green color when cooked and are of a delicious marrow-like flavor, much superior to most early sorts.......................... 10

TREE PUMPKIN OR ZAPPALLITO FROM BRAZIL. This is of a bushy habit of growth and bears its fruit in a cluster near the root of the vine, eight or ten to the plant. Excellent for pies. It resembles the Turban Squash in shape..................... 10

"NEGRO," OR NANTUCKET PUMPKIN. This is the true old-fashioned black-warted shelled pumpkin of old times. The "pumpkin pie" pumpkin of our grandmothers.................. 6

BUTMAN SQUASH. (See page 6.).......................... 10

MARBLEHEAD SQUASH. (See eng. and description, page 7.) 10

CAMBRIDGE MARROW SQUASH. Earlier than the Boston Marrow. The skin has a remarkable deep orange color which makes the squash very attractive to the eye. Popular with market men. Quality hardly up to Boston Marrow.................. 6

YELLOW VICTOR TOMATO. A beautiful golden Tomato, in earliness and shape resembling Canada Victor. Per oz. 40 cts.. 10

LITTLE GEM TOMATO. A prolific variety and desirable for those who wish a small, nice tomato a little larger than the Plum Tomato....................................... 10

CONQUEROR TOMATO. Handsome. Resembling somewhat Canada Victor, but not as large, solid or always early. Vines small... 10

LIVINGSTON'S ACME TOMATO. This is a purple variety having all the fine symmetry and smoothness of the Paragon, differing indeed from it in color mostly.................. 15

POWELL'S TOMATO. Of good size; round, smooth, solid, and ripens well around the stem, and a first-rate bearer. A good variety for market or family use.................... 15

VEGETABLE CATERPILLARS. Large bodied and hairy. These are curious seed-vessels of low growing plants, which strongly suggest the animal caterpillar. Used to decorate side dishes... 10

VEGETABLE SNAILS. Singular seed-vessels of low growing plants, which have a striking resemblance to the snail's of the garden.. 10

WHITLOOF. A most distinct and entirely new vegetable, somewhat resembling Chickory in habit. It produces a moderate-sized and beautiful white heart, in shape similar to Cos Lettuce; the top, either boiled or eaten as a salad, or the root boiled will be found a valuable acquisition................... 10

AGRICULTURAL TREATISES.

A New Treatise.

CARROTS, MANGOLD WURTZELS AND SUGAR BEETS; WHAT KINDS TO RAISE; HOW TO GROW THEM AND HOW TO FEED THEM. BY J. J. H. GREGORY, Marblehead, Mass.
The increased attention given to the raising of roots for feeding to stock, particularly the Carrot and Mangold Wurtzel, has led me to write this treatise. I have endeavored to follow the manner presented in my other works, and give that minuteness of detail in every step of progress, from the seed to the matured crop, that is generally desired by the public. While this work is more particularly intended for persons of limited experience, yet it gathers up so much of experience and observation, covering so much ground in the growing and handling of these two standard crops, that I should be disappointed if about every grower did not find within its covers some facts of more value to him than the cost of the book. Single copies by mail, thirty cents.

CABBAGES, AND HOW TO RAISE THEM. BY JAMES J. H. GREGORY, Marblehead, Mass.
This treatise gives all the minute instructions so valuable to the beginner. It begins with the selecting the ground, and carries the reader along, step by step, through the preparing of the soil, manuring, ploughing, planting, hoeing, weeding, gathering the crop, storing and marketing it, with a hundred minute details embracing every department of the subject.
To prepare myself the more thoroughly to write on this work I experimented on foreign and native varieties of cabbage for four years, raising not far from seventy kinds. The gist of my experience will be found in this treatise. It is illustrated by several fine engravings. I have added a page on the green worm that is causing so much trouble in some localities. Price 30 cents.

ONION RAISING, WHAT KINDS TO RAISE, AND THE WAY TO RAISE THEM. BY J. J. H. GREGORY, Marblehead, Mass.
This work, which I issued in 1865, has been warmly recommended by some of the best authorities in the country, and has gone through fourteen editions. It treats on Onions raised from seed, Potato Onions, Onion Sets, Top Onions, Shallots, and Rareripes, the Onion Maggot, Rust, the merits of the different varieties of Onions, instructions in seed raising, and how to tell good seed,—beginning with the first step of selecting the ground, and carrying the reader along, step by step, through the preparing of the soil, manuring, ploughing, planting, hoeing, weeding, gathering the crop, storing and marketing it, with a hundred minute details embracing eve'y department of the subject.
Illustrated with thirteen engravings of Onions, Sowing Machines and Weeding Machines.
Single copies sent by mail, prepaid, for thirty cents. Seed dealers and booksellers supplied at the usual discount.

SQUASHES AND HOW TO GROW THEM. BY JAMES J. H. GREGORY, Marblehead, Mass.
This treatise is of about the same size and style as my treatise on "Onion Raising," and contains several illustrations, including a section of my squash house, with full directions for erecting one. In plan and thoroughness it is similar to my Onion treatise, very minute and thorough. Beginning with the selection of soil, it treats of the best way of preparing it; the best manures and the way to apply them; planting the seed, protecting the vines from bugs and maggots, the cultivating, gathering, storing and marketing of the crops—giving hundreds of minute details so valuable to inexperienced cultivators. I have written this and my other treatises on the theory that what the public want is minuteness and thoroughness of detail. The price of this is thirty cents, sent by mail post-paid. Dealers supplied at a discount.
If after reading either of these works, any person thinks he has not had his money's worth, let him return them and I will return the money, as I intend that every man shall have his quid pro quo.

Bastian's
Early Blood Turnip Beet.

Bastian's
Half Long Beet.

Long Smooth Beet.

Egyptian Beet.

Early Bassano Beet.

Norbiton's Giant Mangold Wurtzel.

Danvers Early Yellow Onion.

Phinney's Watermelon.

Potato Onions.

Covent Garden Radish.   Large Dutch Parsnip.   Large Red Onion.   Red Turnip Radish.   French Breakfast Radish.

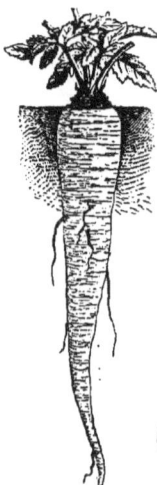

☞ **Pounds, Bushels &c., are priced on these pages instead of separate ones as formerly.** ☜   PRICES OF SEEDS.

ASPARAGUS.

Sow the seed in the seed bed late in the fall or in the early spring, as soon as the ground can be worked, in drills one foot apart, covering the seed about one inch deep. Thin the plants to three inches in the row. The roots may be removed to the permanent bed when one or two years old. In preparing the ground no pains should be spared, as a well established and carefully cultivated asparagus bed will continue in good condition for twenty-five years or more. Select deep, rich, mellow soil, and trench the ground two feet deep, using a liberal quantity of well decomposed manure, with a small admixture of common salt. Set the roots so that the crowns will be three or four inches below the surface of the ground. Apply a dressing of manure in autumn (rotten kelp is excellent), digging the same into the ground in the spring, taking care not to injure the roots. The bed will produce shoots fit for cutting the second or third year after transplanting. An occasional application of salt will be found beneficial.

	lb. exp.	lb. mail	¼ lb.	oz.	pk'g
Defiance (New) See page 14...					
Conover's Colossal. The largest variety grown...	60	75	3.00 20	1.00 10	25 6

BEANS.   Dwarf, Snap or Bush.

Select light, warm soil, and plant when danger from frost is past in the spring, in drills two to two and a half feet apart, dropping the beans about two inches apart in the drill, and cover one inch deep. Keep the ground clean and loose by frequent hoeing, but do not draw the earth around the plants. Avoid working among the vines when they are wet, as it will tend to make them rust.

	peck exp.	b'sh mail	qt. exp.	qt. mail
Chinese. (New.) See engraving elsewhere..package only				15
Dwarf Russian. (New.) See page 14..package only				10
Rose. (New.) See engraving elsewhere..			45	75	15
Dwarf Golden Wax or York Dwarf Wax. (New.) See page 15. More prolific, with larger beans and pods than the common sorts..	1.75	7.00	40	70	10
Early Fejee. Very early, hardy and prolific; on moist garden soil will bear more pickings than any other bean ...	1.50	5.00	25	55	10
Early China, or "Red Eye." An old, popular, early variety..	1.50	4.50	25	55	10
Early Valentine. Pod long, round and tender; excellent, standard early bean in Middle States...........	1.50	5.00	25	55	10
Early Mohawk. Very hardy, early and productive...	1.50	5.00	25	55	10
Early Yellow Six Weeks. Very early and productive; a standard sort..................................	1.50	5.00	25	55	10
Early Rachel. A long, straight-podded, early kind; grown for the New York market....................	1.40	4.00	25	55	10
Dwarf Wax. Pods mostly yellow. Early; for a snap bean, superior....................................	1.75	7.00	35	65	10
Dwarf German Wax. Long variety. Considered more productive than the Round variety.................	1.75	7.00	35	65	10
Dwarf German Wax. Round. A great acquisition; white pods, very early, first rate; beans pure white....	1.75	7.00	35	65	10
Dun Cranberry. One of the very best for stringing; yield first rate; early, good either as a green or dry shell bean ..			25	55	10
Refugee, or Thousand to One. A very prolific bush sort..			25	55	10
Intermediate Horticultural. A half bush variety, very prolific; an excellent substitute for the pole Horticultural; a superior sort for market gardeners.....................................	1.65	6.50	35	65	10
Improved Yellow Eye. One of the best varieties for baking. Remarkably vigorous, healthy and prolific........	1.30	4.00	20	50	6
Concord Bush. A fine, early, bush variety...			35	65	10
True White Pea Bean. (New.) The only strain of Pea Bean that is really round like a pea in shape; fine for baking....................			60	90	10
Navy or Pea Bean. A small, almost round variety, very productive. A standard sort for field culture.......	1.30	4.00	20	50	6
White Medium. White bush variety, largely used by government..	1.30	4.00	20	50	6
White Marrow. A standard sort for field cultivation; early..	1.30	4.00	20	50	6
Red Kidney. A standard red sort..	1.30	4.00	25	55	6

Pole, or Running Varieties.

Set the poles three by three or four feet apart, and plant six to eight beans, with the eyes downward, around each pole, thinning to four healthy plants when they are up. They require the same soil and treatment as the dwarf varieties with the exception that they crave stronger soil, and do best in a sheltered location.

			qt. exp.	qt. mail	pk'g
Golden Butter. (New.) See page 14...			75	1.05	15
Yellow Podded White Wax. (New.) See page 14..			75	1.05	15
Marblehead Champion. (New.) To be used as a string bean. (See page 7.).....................package only					15
Early Lima or Sieva; called also Frost Bean. This is two weeks earlier than Large Lima. Requires the entire season in the North...............			60	90	15
Mottled Cranberry. Long podded, very productive; a very popular bean for garden cultivation............			45	75	10
London Horticultural, or Wren's Egg. Productive, pods elegantly striped—excellent string or shell........	1.75	6.00	40	70	10
Rhode Island Butter. I esteem this bean as one of the very best raised in the North, to eat green-shelled....			40	70	10
Kentucky Wonder. (New.) See page 15..					15
Lamberson's White. (New.) See page 15...package only					15
Large Lima. As a shell bean surpasses all in quality; too late for the extreme North..................	3.00	$10	40	70	10
Dreer's Improved Lima. Rather later but more prolific than Large Lima...............................			50	80	10
Indian Chief or Black Algerian. Always in order for stringing; pods almost transparent, of a yellowish-white...			50	80	10
Boston Market Pole Cranberry. The Boston marketmen cultivate this as the most prolific Pole Cranberry Bean for market...........	2.50		50	80	15
Concord Bean. This new pole bean takes exceedingly well to the poles, is healthy and very prolific; excellent either as string or shell, resembling the Horticultural to which it is related, though it takes better to the poles than that variety, and is considerably earlier.......................	3.00	$10	40	70	10
Caseknife. A white pole bean of great richness either as green shelled or when baked..................			40	70	10
Yard Long. (See page 15.)...					15
Giant Wax. Always a snap bean; a variety that is never stringy at any stage of growth. Pods of a yellowish-white color, very long and remarkably tender. An acquisition...................			60	90	10
White Pole Cranberry. A capital late variety, particularly as a string bean..........................			40	70	10
Painted Lady. Either for ornament or use..package only					15

ENGLISH BEANS.   Broad Windsor.   Large and excellent.

			35	65	10

These thrive best in rich, moist soil and cool situation. Plant in early spring, two or three weeks earlier than the common beans, in rows two feet apart and six inches apart in the row, covering two inches deep. Pinch off the tops of the plants when the young pods first appear.

BEET.

Select a deep, rich, sandy loam, and manure with well decomposed compost. Sow in drills fourteen to sixteen inches apart, and cover one inch deep. When the young plants appear, thin to four or five inches apart. For early use, sow as soon as ground can be worked in the spring; for autumn use, about the middle of May, and for winter use, from the tenth to the twentieth of June, according to variety; the LONG varieties requiring more time to mature than the ROUND, EARLY kinds. When sown late increase the quantity of seed. When young, the plants make excellent "greens". To preserve during winter, cover with earth to keep from wilting. When cooking, boil new beets one hour, and old ones two hours or more. (The Mangold Wurtzels are grown principally for stock, and as they grow larger require more room. They should be sown in drills about two feet apart, and be thinned to twelve or fifteen inches apart in the row. (See my work on Mangold Wurtzels, etc.)

	lb. exp.	lb. mail	¼ lb.	per oz.	
Table Varieties.					
Eclipse. (New.) See page 14..	1.35	1.50	60	25	10
Egyptian. Earlier than Bassano. Tops remarkably small. Excellent for market purposes. I heartily recommend this valuable variety to the attention of market gardeners, who seek above every thing else, earliness	85	1.00	30	12	6

☞ **Pounds, Bushels, &c., are priced on these pages instead of separate ones as formerly.** ✍

	lb. exp. mail	lb.	¼lb.	oz.	pk'g

BEET.

Early Bassano. One of the earliest.
Bastian's Early Blood Turnip. As early as the Bassano, but of a much darker color : excellent every way for early. (See page 15.) — 85 1.00 30 12 6
Early Blood Turnip. A standard sort; good for summer or winter. — 60 75 20 10 6
Hatch's Early Turnip. Somewhat flat in shape; quite a favorite with some of the market gardeners around Boston. — 60 75 20 10 6
Dewing's Early Blood Turnip. Very symmetrical; free from fibrous roots; dark red. This has taken several first premiums at the Massachusetts State Fair. — 60 75 20 10 6
Simon's Early Turnip. About as early as Early Bassano, but of a deeper red. Popular in the Philadelphia market. — 60 75 20 10 6
Yellow Turnip. A very early sort, about as early as Bassano; of a beautiful golden yellow color. — 85 1.00 30 10 6
Bastian's Half Long Blood. A new Philadelphia sort of a fine dark color; a good grower. — 85 1.00 30 12 6
Henderson's Pine Apple. Excellent for family use. — 85 1.00 30 12 6
Long Smooth Dark Blood. Excellent for winter use; smooth skinned; flesh dark red. — 60 75 20 10 6
Dell's Ornamental Dwarf. This has leaves of a peculiarly deep, rich red color, and is cultivated in Europe as an ornament in the flower garden; grows partly above ground. ...package only — 10

Mangold Wurtzels, Varieties for Feeding Stock and for Sugar Making.

Knauer's Improved Imperial. New. A standard German variety for making sugar. — 85 1.00 30 12 6
Improved American Sugar or Lane's. A long white variety of Mangold Wurtzel, for stock. — 65 80 25 10 6
Vilmorin's Improved French White Sugar. This is the variety cultivated by the French for the manufacture of sugar. Of six varieties of beets tested for sugar at the Farm of the Maine Agricultural College last season my seed of this variety gave the highest per cent. of sugar. — 85 1.00 30 12 6
Carter's Orange Globe Mangold Wurtzel. The best variety of Yellow Globe. — 60 75 20 10 6
Carter's Mammoth Mangold Wurtzel. Said to excel in size. — 60 75 20 10 6
Yellow Ovoid Mangold Wurtzel. The Ovoid Mangolds grow more symmetrical and freer of rootlets than the long sorts. They are heavier, bulk for bulk. — 60 75 20 10 6
Red Giant Ovoid Mangold Wurtzel. Very large, oval shape; pulls up very free from dirt. — 60 75 20 10 6
Norbiton Giant Mangold Wurtzel. A new English variety which tends less to a hollow neck than the old Long Red, kind. — 60 75 20 10 6

Red Globe Mangold Wurtzel.
White Sugar.
Yellow Globe Mangold Wurtzel.
} The Globe Mangolds succeed better than the long sorts on sandy soil. All the varieties of Mangolds are excellent food for cows, to increase the flow of milk. Farmers should begin to feed them towards the close of winter and in the spring. — 40 55 15 10 6

BERBERRY. — 1.10 1.25 40 15 8
One of the best shrubs for hedges. Perfectly hardy. Never winter-kills and grows on any soil; makes a thick, close, impenetrable hedge that will turn cattle, and promises to become the hedge plant of North America. The berries make excellent preserves. As the seeds of the Berberry do best when planted in the berry, I will receive orders and fill them to be filled in the fall, as soon as the fruit is matured, when I will send the berries at prices named above with a page of full directions for making a hedge, and for preserving the fruit.

BORAGE. — 20 6
This is a profuse flowering plant, which is grown principally for bees, or as ornament in the flower garden. Sow in early spring in rich soil, and thin plants to one foot apart. It readily bears transplanting, and when thus treated flowers more abundantly.

BRUSSELS SPROUTS
A class of plants allied to the Cabbage family, producing great numbers of small heads or sprouts on the main stem of the plant, which are used in the manner of Cabbages. Plant in rich soil in hills two feet apart each way, and thin to one plant to the hill.
Scrymger's Giant Dwarf. (New.) In habit close headed and compact. ...package only — 10
Dwarf Improved. — 20 6
Dalmeny Sprouts. A hybrid between Drumhead Savoy and Brussels Sprouts. — 25 10

BROCCOLI.
The Broccoli are closely allied to the Cauliflower family, so nearly so that the Walcheren variety is sometimes classed with Cauliflower. They require similar cultivation and treatment to Cauliflower.
Walcheren White. One of the very best varieties. — 75 10
Large White Early French. A standard French variety. — 50 10
Knight's Protecting. Dwarf, very hardy; heads very large for the plants. — 50 10
Purple Cape. Late, large, compact. — 50 10
Early Purple. Early, excellent; color deep purple. — 40 10
Elletson's Mammoth. A large English variety. — 50 10

CABBAGE.
Cabbage will thrive on any good corn land, though the stronger the soil the better they will develop. New land is preferable. Plough deep and manure very liberally. The early sorts bear planting from eighteen inches to two feet apart in the rows, with the rows from two to two and a half feet apart. The large varieties to be from two to four feet apart in the rows, with the rows from two and a half to four feet apart. The distance varying with the size. The crop should receive as many as three hoeings and three cultivatings. Cabbage will not usually follow cabbage or turnips successfully in field culture, unless three or four years have intervened between the crops. For late fall marketing, plant drumhead sorts from June 10th to 20th For full and minute information in every department of Cabbage culture, see my treatise on "Cabbages, and How to grow them."

Earliest Varieties.

	lb. exp. mail	lb.	¼lb.	oz.	pk'g
Vilmorin's Early Flat Dutch. (New. See page 15.) Heads rounder and harder than the common variety.	3.85	4.00	1.30	40	10
Early Bloisfeld Giant. (New. See engraving elsewhere.	4.85	5.00	1.50	50	10
Heartwell Early Marrow. New. (See page 15.) ...package only					15
Henderson's Early Summer. (New. See engraving elsewhere.	6.75	7.00	2.00	70	15
Crane's Early. (New.) (For des. see page 15.)				50	15
Early Nonpareil. A choice very early sort.	1.35	1.50	50	15	6
Carter's Little Pixie Savoy. This variety closely resembles in earliness and size the Little Pixie.				25	6
Wheeler's Cocoanut Cabbage. A new sort, conical in shape, making fine hard heads; one of the best of the English early market varieties.				40	10
Early York. One of the earliest; an old standard sort.	1.35	1.50	50	20	6
Large York. An improvement in size on Early York; a little later.				20	6
Early Jersey Wakefield. (True.) Resembles Oxheart. A standard early cabbage in Boston and New York markets.	4.85	5.00	1.50	50	10
Little Pixie. A small, very tender and sweet cabbage, of the pointed heading family. It is earlier than Early York and heads hard, and from its small size a great number can be matured on a small area of land.	2.35	2.50	75	25	10
Sugar Loaf. A popular early variety.				25	6
Early Oxheart. An excellent early sort.				25	6
Early Wyman. This new cabbage was originated by Captain Wyman, of Cambridge, Mass. It is allied to the early Wakefield, is about as early but grows to double the size; very popular with market gardeners as an early market sort.	3.85	4.00	1.50	40	10

☞ Pounds, Bushels, &c., are priced on these pages instead of separate ones as formerly. ☜ | PRICES OF SEEDS.

CABBAGE.

	lb. (exp. mail)	lb.	¼ lb.	oz.	pk'g
Cannon Ball. The hardest heading of all early sorts.	2.85	3.00	1.00	30	10
Early Ulm Savoy. One of the earliest; unsurpassed in quality; capital for family use.				40	10

Second Early.

"Newark" Early Flat Dutch. The best strain of second early variety in the New York market; heads large, solid, broad and thick.	3.85	1.00	1.30	40	10
Early Blood Red Erfurt. (New. See page 15.) Heads darker red than common sorts.............package only					25
Fottler's Improved Early Brunswick. The earliest of the large heading drumheads. This has given great satisfaction in every section of the United States. (See page 10.).	3.85	3.00	1.30	40	10
Early Winnigstadt. Heads large, cone-shaped and solid; one of the very best for all soils.	2.35	2.50	75	30	10
Large French Oxheart. Popular as an early cabbage.				30	6
Schweinfurt Quintal. The earliest of all large drumheads; grows from a foot to eighteen inches in diameter; does not head very hard, but is remarkably tender. The heads are very handsome, and almost as rich as the Savoy class.	3.85	4.00	1.30	40	10
Early Red Erfurt. Early, head round and very solid.				40	10

Late Kinds.

Marblehead Dutch. (New. See page 15.).	3.85	4.00	1.30	40	10
Improved American Savoy, Extra Curled. Very reliable for heading; more finely curled than Improved American Savoy, which renders it very desirable for market gardeners and for family use.			1.30	40	10
St. Dennis Cabbage. A large late drumhead, makes a very solid head. Popular in Canada.				35	10
Green Glazed. A standard variety in the South.				35	10
✗ **Marblehead Mammoth Drumhead.** The largest cabbage in the world. (See page 10.).	4.85	3.00	1.50	50	10
Bergen Drumhead. A standard in New York market.	2.35	2.50	75	25	10
Stone Mason Drumhead. A standard variety in Boston market. (See page 10.).	3.85	4.00	1.25	40	10
Premium Flat Dutch. Large and excellent for winter; very extensively grown.	2.35	2.50	75	30	10
Improved American Savoy. An improvement on the old Green Globe Savoy; very reliable for heading. Very sweet and tender—much esteemed for family use. An excellent sort for market gardeners.	2.85	3.00	1.00	30	10
Drumhead Savoy. A cross between Savoy and Drumhead—very large.	2.35	2.50	75	25	10
Red Dutch. The old variety for pickling.	2.35	2.50	75	30	6
Red Drumhead. Larger than Red Dutch and more profitable; heads round; very reliable for heading, very hard under high cultivation.	2.85	3.00	1.00	35	10

CARROT.

Carrots thrive best in rather a light loam. The ground should be well manured with fine, well rotted or composted manure, six or eight cords to the acre, and be thoroughly worked quite deep, by two ploughings made at right angles with each other. Also cultivate and drag if there are any lumps, and then rake level, burying all remaining lumps and stones. Plant in rows fourteen inches apart, and thin plants three to five inches in the rows. Plant from the middle of April to middle of May, to insure crop; though good success is often met with if planted as late as 10th of June. As the dry spells which sometimes prevail at that season are apt either to prevent the germination of the seed, or to burn the plants as soon as they appear above ground, it is therefore advisable to increase the quantity of seed, which under the circumstances will give the crop a better chance. Keep very clean of weeds. (See my work on Mangolds and Carrots, page 17.)

Danvers. (New.) (See page 8.).	1.35	1.50	50	15	6
Early Very Short Scarlet. The earliest and smallest of all varieties; of special value for forcing.				15	6
Early Scarlet Horn. The early short variety for forcing; excellent for the table; color very deep orange.				15	6
✗ **Short Horn.** The standard early variety; sweeter than Long Orange and more solid. Good to color butter.	1.05	1.20	40	15	6
Improved Long Orange. Of a darker, richer color than Long Orange.	85	1.00	30	12	6
Long Orange. The standard field carrot; good for stock.	85	1.00	30	12	6
Large Altringham. Bright orange; grows a little above ground. A poor cropper.	85	1.00	30	12	6
Large White Belgian. Largest of all, white and most productive; good for horses; entire crop can be pulled by hand.	60	75	20	10	6
Yellow Belgian. Grows partly out of ground. A capital sort for late keeping.	85	1.00	30	12	6

CAULIFLOWER.

Pursue the same course as with Cabbage, manuring rather heavier and hoeing oftener. Cauliflowers covet the cool, moist weather of the fall months to perfect themselves.

Gerry Island. (New.) See engraving elsewhere.				3.00	30
Berlin Dwarf. (New.) See page 14.		5.00	1.30	25	
Late Algerian. (New.) See page 14.		3.00	1.00	20	
Henderson's Early Snowball. (New.) See page 15. Very dwarf; very early; very reliable. Price per ½ oz. 2.50				8.00	50
Autumnal Late Giant. (New.) Very large headed and extremely productive.............package only					25
Italian Early Giant. Fine, large, white-headed and early.		5.00	1.30	25	
Carter's Dwarf Mammoth. A premium English variety; very early, with heads remarkably large for so dwarf a variety.				1.50	25
Dwarf Early La Maitre. A new French sort, making fine large heads.				1.50	15
Early Paris. A standard early variety.	9.85	810	3.50	1.00	15
Early Erfurt. A choice German variety.		5.00	1.50	15	
Extra Early Dwarf Erfurt. Extra choice. Specially selected. (See page 15).				4.00	50
Fitch's Early London. The best strain of this standard English sort.				75	15
Early Dutch. Early.				1.00	15
Nonpareil. One of the earliest varieties, resembles Improved Early Paris.				1.00	15
Lenormand's Short-Stemmed Mammoth. Dwarf, large and fine. One of the largest and the most reliable for general cultivation.............per pound by express 813.85	814.	4.50	1.50	25	
Large White French. Fine, large white.				75	15
Stadtholder. Fine; large size; late.				1.00	25

CELERY.

Plant seed in hot bed or very early in open ground. Transplant four inches apart, when three inches high, in rich soil finely pulverized; water and protect until well rooted, then transplant into rows five or six feet apart either on surface, or in well manured trenches a foot in depth, half filled with well rotted manure. Set the plants from eight to twelve inches apart. To blanch draw earth around the plants from time to time, taking care not to cover the tops of the center shoots.

Crawford's Half Dwarf. (New. See page 4.).	3.85	4.00	1.25	40	10
Sandringham Dwarf White. Most dwarf of all; very solid; white.	2.35	2.50	75	30	10
White Solid. A standard sort.	2.35	2.50	75	25	6
Boston Market. Short, compact and solid—very popular; almost the only variety sold in the Boston market. (See page 15.).	3.85	4.00	1.25	40	10
Turnip Rooted. The root of this is eaten.	2.35	2.50	75	25	6
Carter's Crimson. Dwarf, solid and crisp; a first class variety.				30	10
Turner's Incomparable Dwarf White Solid. Popular in England, and extensively grown by the New York market men. In dwarf habit next to Sandringham.	2.35	2.50	75	25	6

[☞ Pounds, Bushels, &c., are priced on these pages instead of separate ones as formerly. ☜]   PRICES OF SEEDS.

CHICKORY.

Pursue the same manner of cultivation as for Carrot. If to be used as a salad, blanch the leaves by covering so as to exclude the light. If raised for its root, dig at about the same time as Carrots, wash the roots and then slice them, either way, and dry thoroughly by artificial heat.

	lb. exp.	lb. mail	¼lb.	oz.	pk'g
Large Coffee Rooted. Used as a substitute for coffee	85	1.00	30	20	6
CHUFAS, or Earth Almonds. (See page 15.)	85	1.00	30	15	6
COLLARDS or COLEWORTS. True Southern.				20	6

A class of plants closely allied to the Cabbage family, which are somewhat extensively used in the South, when small, as greens. Sow in early spring in drills one foot apart, covering the seed half an inch. The young plants are ready for use as soon as they have attained sufficient size, but if it is desired to keep them in good condition, thin the plants to six or eight inches apart, and pull off the larger leaves before using.

CORN.

Corn revels in a warm and rich soil. Do not plant before the ground has become warm—nothing is gained by it. Drill cultivation is more profitable than hill cultivation. The smaller varieties may be planted with the drills two and a half feet apart, and the stalks thinned to ten inches apart; the larger sorts should have the drills three to four feet apart, and the stalks a foot apart in the rows, and the largest varieties eighteen inches apart. Use some rich manure in the drills. Frequently stir the earth around the roots by hoe or cultivator, but do not draw it up about the stalks. For a succession of corn for family use to be planted at the same time, I would recommend Marblehead Early, Pratt's, Crosby's, Moore's, Stowell's and Egyptian Sweet.

Sweet Varieties for Family use and Marketing in a green state.

	pork b'sh exp.	qt. exp.	qt. mail
Early Boynton Sweet. (New.) See page 14.................package only			10
Marblehead Early Sweet. The earliest of all; allied to the Narragansett but a week earlier. See page 4	55	75	15
Egyptian Sweet. New. (See page 6.)	1.50 5.00	43	65	10
Forty Days. Earlier than our standard early corn, and will be found desirable in Northern latitudes as an extremely early sort for a flint variety	40	60	10
Pratt's Early. Here we have a capital sort for marketmen who are looking about for an early sort, growing to a fair market size. (See page 16.)	1.50 5.00	30	50	10
Early Minnesota Sweet. One of the very earliest sorts of sweet corn, with ears of suitable size for market purposes.	1.25 4.50	30	50	10
Early Narragansett. One of the earliest; kernels very large; ears large in diameter, and of medium length.	1.25 4.50	30	50	10
Moore's Early Concord Sweet. A new early corn, from 12 to 16 rows. Remarkably handsome; quite popular. Awarded a silver medal by the Mass. Horticultural Society	1.25 4.50	30	50	10
Crosby's New Early Sweet. First rate every way, either for market or family use.	1.25 4.50	30	50	10
Mexican Sweet. The sweetest and tenderest for table use of all varieties I am acquainted with.	1.25 4.50	33	55	10
Golden Sweet. The only cross ever made between the sweet and field varieties; flavor, peculiarly rich	35	55	10
Stowell's Evergreen Sweet. Excellent; keeps green till cold weather; ears large; a standard late variety.	1.25 4.00	35	55	10
Gen. Grant. An acquisition because of its extreme sweetness. Late, comes in after Stowell's. The best for fodder.	1.25 4.50	35	55	10
Marblehead Mammoth Sweet. The largest variety grown. (See page 16.)	1.50 5.00	45	65	15
Sweet Fodder Corn. Sweet corn is preferred to the yellow kinds by our best farmers for fodder.	1.00 3.00	25	45	6

Varieties for Field Cultivation and Popping.

Blunt's Prolific Field. (New.) See engraving elsewhere.	1.40 4.00	40	60	10
Longfellow's Field. (New.) See page 8.)	1.60 3.00	35	55	10
Adams' Early. A favorite in the South. The earliest of all the Dent sorts.	1.30 4.00	30	50	6
Improved Early Yellow Canada. A first rate corn where the seasons are short.	1.00 3.00	30	50	6
Lamson's Early Yellow Field. An excellent variety for latitude of New England; ears quite large and well filled out; two hundred and fourteen bushels of ears have been grown on an acre	1.00 3.00	30	50	6
Hundred Days Dent. Early; ears large and well filled. Capital for the Middle and in favorable seasons for Southern New England states. Will ripen in one hundred days in a good corn season.	1.25 4.00	25	45	6
Mammoth Field. A large white gourd seed variety, claimed to be the largest kind raised in the Western States.	1.25 4.00	30	50	6
Silver Laced Pop. The handsomest of all varieties of pop corn, and decidedly a growing favorite.	35	55	10
Nonpareil, or Pop. The popular variety for packing.	1.50 5.00	30	50	10
Dwarf Golden Pop. Small, but ornamental, and a favorite with the little folks; excellent for popping.	30	50	10
Egyptian Pop. Tenderer when popped than the common variety.	70	90	15

BROOM CORN.

Any good corn land will grow Broom Corn. Plant in rows three feet apart and thin to eight inches in the row.

California or Golden. (New.) See page 14.	55	75	15
Improved Evergreen. An improvement on the Evergreen by careful selection of stock for years. Not as tall as Evergreen; brush fine and bright colored.	1.50 5.00	40	60	10

CRESS.

Plant on rich soil, finely pulverized, in drills six or eight inches apart. That grown in the cool of the season is of the best quality. To be used as salad before the flowers appear.

	lb. exp.	lb. mail	¼lb.	oz.	
Curled. The best sort.	45	60	20	10	6
Plain or Common.				10	6
Water Cress. To be planted along the borders of shallow water courses. The famous English Cress.				10	

CUCUMBER.

The vines require a warm location. Plant after the ground has become warm, in hills four feet apart for the smaller varieties, and five feet for the larger sorts. Manure with ashes, guano, or some well rotted compost, working the manure just under the surface. Sprinkle vines with plaster or air-slacked lime to protect it from bugs. The frame cucumbers can be successfully grown in the open air in this country by giving them well sheltered location, plenty of manure, and having hills six by six.

	lb. exp.	lb. mail	¼lb.	oz.	
White German. (New.) See engraving elsewhere.					25
Extra Long Green Smooth. (New.) See page 15. Very long, smooth and straight..........package only					15
Long Green Smooth from Athens. (New.) See page 15					30
Short French Pickling. (New.) See page 14..........package only					15
Marquis of Lorne. (New.) A celebrated frame variety, short neck, smooth skin; very straight and prolific, package only	1.25 1.50	50	20		50
Green Prolific. (New.) See page 15.	1.25 1.50	50	30		10
Bismarck. (New.) (See page 15.)				30	15
Chinese Long Netted. (New.) See page 16..........package only					25
Russian Netted. (New.) (For description see page 16.)				30	10
Tailby's Hybrid. (New.) (For description see page 13.)				30	10
Rollisson's Telegraph. (New.) One of the most prolific of the forcing varieties.	1.35 1.50	50	30		20
Norbiton Giant. (See page 16.) The longest prize frame cucumber known..........package only					20
English Prize Cucumbers. Carter's Champion; Sion House. These yield but very few seed, and are great favorites in England. Each variety per package					25
Gen. Grant. The hardiest and probably the most prolific of the English Frame varieties. In England the climate is not hot enough to grow cucumbers in the open air.					20
Early Russian. The earliest of all varieties; grows about four inches long.	1.10 1.25	40	12		6

☞ Pounds, Bushels, &c., are priced on these pages instead of separate ones as formerly. ☞   PRICES OF SEEDS.

CUCUMBER.

	lb. exp.	lb. mail	¼ lb.	oz.	pk'g
Early Cluster. Bears mostly in clusters; very early and productive	85	1.00	30	12	6
Improved White Spine. Great bearer; excellent for early forcing, or for out door cultivation; standard in Boston market. My stock is from one of the best Boston market gardeners	85	1.00	30	12	6
✓**Early Frame.** Early, short, prolific	85	1.00	30	12	6
Long Green. An old standard sort	85	1.00	30	12	6
Short Green. An old standard	85	1.00	30	12	6
Improved Long Green Prickly. Excellent variety, growing 18 or 20 inches long; makes a hard brittle pickle				25	6
West India Gherkin. A very small, elegant, peculiar sort, for pickles only, prolific to an extraordinary degree. Somewhat difficult to get the seed to germinate				30	16
✗ **New Jersey Hybrid.** The largest of all white spined varieties	1.35	1.50	50	15	6
Eight Varieties Mixed				20	6
Boston Pickling. A medium, long variety; the standard for pickling in Boston market	1.10	1.25	35	15	5
Short Pickling. Very desirable for a short pickle	85	1.00	30	15	6
Early White Japan. A variety recently introduced from Japan, exceedingly productive; resembles White Spine, but turns to a richer creamy white color, and is earlier	1.10	1.25	35	15	6
Snake. I have grown these six feet in length, coiled up like a snake. (See page 16) ....package only					20

DANDELION.

This vegetable has become very popular as an early beautiful green, and the roots also are used when dried as a substitute for coffee. Its use in either of these forms is particularly recommended to those who are inclined to any disease of the liver. Sow in May in drills one foot apart, covering the seed half inch deep. A rich soil is preferable, but this plant will thrive anywhere

New Very Double. New. See page 15 ....package only					15
Improved Thick Leaved. (New.) Seven hundred bushels of this sort have been grown on three-fourths of an acre	4.85	5.00	1.70	60	15
Common	3.85	4.00	1.30	40	10

EGG PLANT.

Plant the seed in March, in a hot-bed, or, for family use, in flower pots, in a warm window. Transplant in open ground after warm weather has become warm and settled, in rows two feet apart each way. They require a rich soil and as favorable a location for warmth as the garden will afford.

Long White China. A very delicate and beautiful long white variety. Highly esteemed by amateurs.. package only					15
Very Early Dwarf. A new French variety of Long Purple; extra early. (See page 16) ....package only					15
Striped Guadaloupe. Long in shape and elegantly striped ; very ornamental and edible ....package only					15
Black Pekin. A new variety of Round Purple. Blackish violet leaves; fruit very large				75	15
Long Purple. Earlier and more productive, but smaller than Round Purple				50	15
New York Improved Round Purple. An excellent variety, surpassing in size of fruit				50	15
Scarlet China. (New. A fine ornamental variety					15

ENDIVE.

For early use sow as soon as the ground can be worked in the spring, in drills fifteen inches apart, and thin plants to six or eight inches in the row. A succession may be obtained by sowing every two or three weeks until midsummer, when it will be proper time to plant for fall and winter use. Any common garden soil will do, but a rather moist situation is preferable. To blanch the leaves gather them carefully together when perfectly dry and tie with matting or any soft fibrous material. Another method is to invert flower pots over the plants. The leaves are very highly esteemed for use as salads.

London Green Curled. Very popular				25	6
Fine Curled Mossy. Very ornamental				30	10
Broad Leaved Batavian. A large summer variety				25	10

GOURDS.

The larger varieties require the entire season to mature them, and the ornamental sorts are apt to grow too large if the ground is very rich.

Hercules Club. Grows 4 to 6 feet in length ....package only					10
Sugar Trough. (See page 16) ....package only					12
Double Bottle. ....package only					10
Dipper. Used as its name indicates ....package only					10
Angora ....package only					10
Dish-Cloth Gourd. (See page 16.) ....package only					15
Fancy and Ornamental. Mixed varieties. (See page 16) ....package only					10

GARLICS.

Plant the bulbs on exceedingly rich soil, in rows or in ridges fourteen inches apart and six inches apart in the rows. They are cultivated for their flavor, (which is similar to the onion but more powerful,) and are used in stews, soups, &c.

		25	40	15	10

KALE, or BORECOLE.

Plant the larger sorts in hills two by three feet apart, and thin to one plant to the hill. Select deep, rich soil, and cultivate as Cabbage. Some of the varieties are very ornamental, and scattered singly are attractive in the flower garden, being finely curled and variegated with green, yellowish white, bright red and purple leaves. The tender leaves are used as Cabbage.

Frisby's Crested. (New.) ....package only					15
Green Curled Tall Scotch. (New. See plate on page 34) One of the best varieties	85	1.00	40	20	10
Sea Kale. The young shoots when blanched are exceedingly delicate, being much superior to Broccoli ...package only					10
Carter's Garnishing. Both ornamental and useful. The seed will produce many varieties of high colored plants package only					15
Ornamental Kale, four elegant varieties. For ornament of the table ....package only					15
Dwarf Green Curled, or German Greens. Very hardy ; a standard market sort	85	1.00	40	15	6
Cottager's. A new English variety				20	6
Field Kale. For cattle. Can be cut several times during the season				30	10
Abergeldie. A new dwarf variety, curled as fine as parsley ; of delicate, mellow flavor				50	10

KOHL RABI, or TURNIP CABBAGE.

Prepare ground as for Cabbage, then plant about the first of June in rows two feet apart, thinning plants to twelve inches in the row. To preserve over winter treat as turnips. When young their flesh is tender and resembles a fine ruta baga with less of a turnip flavor. When fully matured they are excellent for stock.

Early White Vienna. A standard early kind	2.85	3.00	1.00	30	8
Large Purple. Very large, hardy and productive ; for stock	2.85	3.00	1.00	30	6

LEEK.

Select good onion soil, manure liberally, and plant in April in drills made six or eight inches deep and eighteen inches apart, and thin to nine inches apart in the drill. Gradually draw the earth around the plants until the drills are filled level with the surface. Draw for use in October. To be used in soups or boiled as asparagus.

Large Musselburg Leek.				40	10
Broad Scotch, or Flag. A large and strong plant ; hardy ; color deeper than Rouen				25	6

☞ Pounds, Bushels, &c., are priced on these pages instead of separate ones as formerly.   PRICES OF SEEDS.

	lb. exp.	lb. small	½ lb.	oz.	pk'g
LEEK.					
Very Large Rouen. A new French variety; best of all for forcing..................................				30	6
Extra Large Carentan. (New.) A very fine extra large winter variety.....................package only					15
LETTUCE.					
Lettuce covers a rich and rather moist soil. The rows should be about twelve inches apart and the plants thinned from eight to twelve inches apart for the heading varieties. When heads are not desired it may be grown in a mass. The more rapid the growth the better the quality. Some varieties are peculiarly adapted for early culture, others for summer growth.					
Nellis' Perpetual. (New) See page 14..................................package only					15
Stone Head Golden Yellow. (New.) See page 16. I invite gardeners to test this for quality with the best variety they know of, believing that this will bear the palm..........................\					15
Satisfaction. New. (See page 16.)...				40	15
Egyptian. (New. See page 16.)..................................package only					15
All-the-Year-Round. New. (See page 16.)..				40	15
Black Seeded Tennis Ball. Hardy; excellent for early crops; earlier than Silesia; large heads. My stocks of this and White Tennis Ball are from one of the first Boston market gardeners. A favorite in Boston market	2.60	2.75	80	30	6
Hanson Lettuce. (See page 13.) Stock of this very large this season.....................	2.85	3.00	1.00	40	15
White Tennis Ball, or Boston. (White seed.) The variety so extensively grown by the Boston marketmen during winter for marketing in February and March. A fine early sort—small heads, very hardy; used for winter culture..	2.60	2.75	80	30	6
Early Curled Simpson. Resembles Silesia, but is more curled and not so early....................	2.10	2.25	70	25	6
Early Curled Silesia. Very early; very tender and sweet—a popular variety for hot-beds and early out-door culture.	1.60	1.75	55	20	6
Early Butter Head. An excellent sort—a great favorite..........................	2.35	2.50	75	30	6
True Boston Curled. The most elegant Lettuce of all. Quality good; very popular................	2.35	2.50	75	30	10
Drumhead. Very large; heads crisp and tender. A standard sort......................	1.35	1.50	50	20	6
Large India. Resembles Drumhead but later; of fine quality......................	3.35	3.50	1.00	35	10
Brown Genoa Cabbage. Of medium size, round head stained with red about the top. One of the best for either summer or winter use..				35	6
Improved Spotted Cabbaging. A fine head variety; color green shaded with brown; quality first rate; one of the finest for the table................................package only					10
Large Princess Head. A new, very fine German variety, which does finely in the United States............				35	6
Perpignan. Heads sometimes seven inches in diameter. One of the best summer varieties. Not inclined to go to seed	2.85	3.00	1.00	35	6
French Imperial Cabbage. A fine large-headed variety; one of the very best for family use, as it does not run quick to seed......................................				35	6
Versailles Cabbage. A fine summer variety; light green; makes large heads......................				35	6
Bossin. A new French variety; large; late; color dark green......................				35	6
Neapolitan Cabbage. A good summer variety..........................	2.85	3.00	1.00	35	6
Six Choice Varieties. Mixed in one package.............................package only					15
White Paris Cos. Best of all the Cos varieties..........................				35	6
Kingholm Cos. Stands the summer heat splendidly and heads without tying; makes fine large heads..package only					10
Green 'Fat' Cabbage. A fine summer cabbage variety; dark green; does not run to seed early....package only					15
Victoria Cabbage. One of the best English cabbage varieties.........................package only					10
MARTYNIA.					
Plant on any rich, garden soil, two by three feet apart, leaving only one plant in a place. It produces an abundance of large, showy flowers, and the young pods, when sufficiently tender to be easily punctured by the nail, are used for pickles.				25	6
MELON.					
Select warm and light soil—a poor light soil is better than a cold and rich one. Thoroughly work the soil, manure with guano, phosphate or a rich compost, having the hills six feet apart for the musk varieties, and eight or nine for water melons. Do not excavate hills, but work the manure just under the surface, as the roots of all vines naturally seek warmth. Pinch the more vigorous vines from time to time, and work in guano or phosphate between the rows. Plant a dozen or more seeds in each hill, but do not leave over two plants. Sprinkle young plants liberally with plaster or air-slacked lime to protect from depredation of insects.					

Musk Varieties.

	lb. exp.	lb. small	½ lb.	oz.	pk'g
Persian. (New.) See page 14................................package only					10
Bay View. (New.) See engraving elsewhere..........................				30	15
Chicago Nutmeg. (New.) See page 14..........................				30	15
Surprise. (New) See page 15.............................package only					15
Log-of-wood. New (See page 8).............................package only					15
Algiers Cantaloupe. (New) See page 15..........................					15
Silver Netted. (New. See page 16)..........................				30	10
Christiana. (True) Remarkable for early maturity..........................	1.10	1.25	40	15	6
Improved Cantaloupe. (New.) A very early, large round sort; first rate for market.......	1.10	1.25	40	20	6
Hardy Ridge. (New) (See page 13) A remarkably thick fleshed melon, of good quality. A most vigorous grower...				40	15
Sill's Hybrid. (True) Salmon colored, flesh rich, sweet and delicious. (See page 16.)..........	1.10	1.25	40	20	6
Torrey's. Green fleshed; large; earlier than Casaba..........................				20	6
Shaw's Golden Superb. Though small in size, superb in quality.—good for family use.				20	6
Skillman's Fine Netted. Of delicious flavor—early..........................	95	1.10	35	15	6
Early Nutmeg. Green fleshed, highly scented; mine is the Boston variety, which is earlier than the Nutmeg grown further South..........................	1.00	1.15	35	15	6
Long Yellow. Large, sweet, productive; a well known sort..........................	85	1.00	30	15	6
Green Citron. Green fleshed; sweet, melting, and rich flavored..........................	95	1.10	35	15	6
Ward's Nectar. Early, exceedingly prolific, sweet, rich, and delicious; green fleshed..........	1.35	1.50	50	15	6
Early Jenny Lind. An early sort; favorite with gardeners..........................	95	1.10	35	15	6
New White Japan. Flesh greenish white; early and prolific; sweet, delicious..........	1.50	1.25	40	15	6
Pine Apple. Oval shaped, rough netted, thick fleshed, juicy and sweet..........................	95	1.10	35	15	6
Casaba. (New.) A very large, long, green-fleshed melon, of delicate flavor, thick fleshed, melting and delicious; has been grown to weigh 15 lbs. An acquisition..........................	1.10	1.25	40	20	10

Watermelon.

	lb. exp.	lb. small	½ lb.	oz.	pk'g
Golden Fleshed. New. (See page 16)..........................				30	15
Excelsior. (New. See page 8)..........................	1.35	1.50	50	25	10
Ferry's Peerless. (New. See page 4)..........................	1.10	1.25	40	15	10
Vick's Early. (New. See page 8)..........................	1.10	1.25	40	20	10
Ice Cream, true, White Seeded. A very early melon of superior quality. Very popular..........	1.15	1.30	40	15	6
Jackson, or Strawberry. New; delicious A great favorite in the Middle and Southern States Seeds white tipped with red..........................	1.35	1.50	50	20	6
Gipsey. Very large and very productive. The principal variety grown in New Jersey and sent to Northern markets.	95	1.10	35	15	6

1   2   3   4   5   6

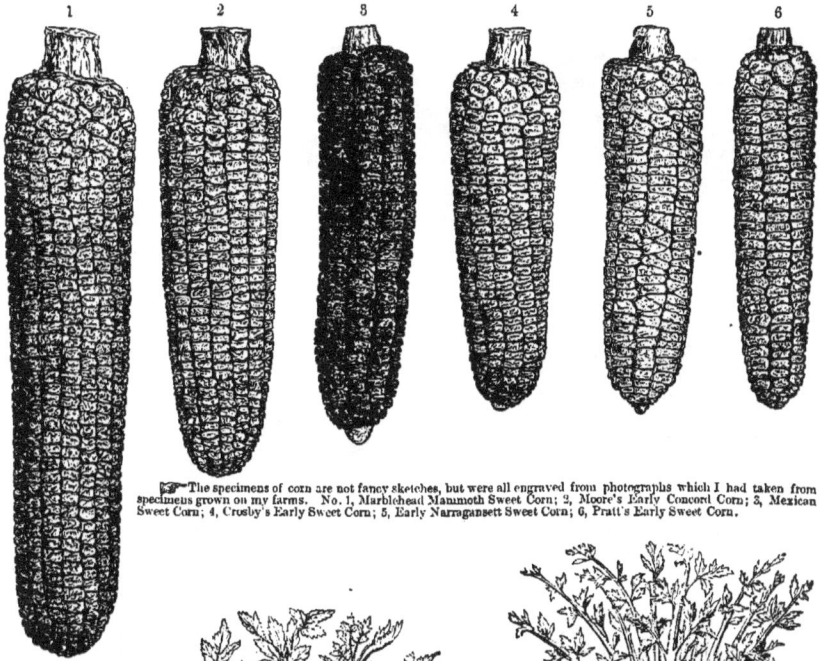

☞ The specimens of corn are not fancy sketches, but were all engraved from photographs which I had taken from specimens grown on my farms. No. 1, Marblehead Mammoth Sweet Corn; 2, Moore's Early Concord Corn; 3, Mexican Sweet Corn; 4, Crosby's Early Sweet Corn; 5, Early Narragansett Sweet Corn; 6, Pratt's Early Sweet Corn.

Improved Large
Yellow, or Cracker
Onion.

Boston Market Celery.

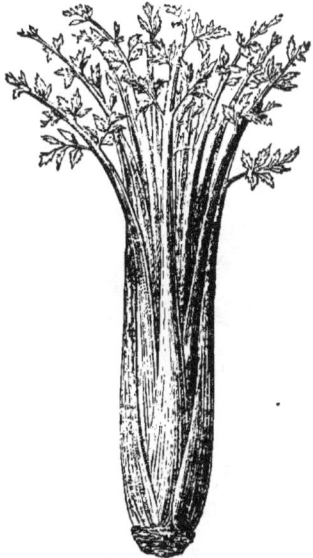

White Solid Celery.

Deep Scarlet Olive-Shaped
Radish.

Beginning at the left hand, the smallest Cucumber is the **Early Russian**
then follow Short Horn, Early Cluster (two specimens, White Improved
Spine, Long Green, and Improved Long Green Prickly.

Chinese Rose Winter
Radish.

⅕

Casaba Muskmelon,

Nutmeg Melon.

Sill's Hybrid Muskmelon.

Russian Netted Cucumber.

Bell Pepper.

Crookneck Squash.

Drumhead Lettuce.

Egg Plant.

Cayenne Pepper.

☞ Seeds per Express or Freight at purchaser's expense. ✍

PRICES OF SEEDS.

MELON. Watermelon.

	lb. exp.	lb. mail	¼lb.	oz.	pk'g
Cream Fleshed Sculptured Seeded. (See page 11) New ; early, remarkably sweet, with seed singularly marked..	1.35	1.50	50	15	6
Mountain Sweet. An old standard variety; early, solid, sweet and delicious; one of the best for northern cultivation.	85	1.00	30	10	6
Mountain Sprout. Long, striped, scarlet flesh ; an old standard sort...	85	1.00	30	10	6
Phinney's. For hardiness, vigor, and productiveness, unexcelled ; early, very reliable, red-fleshed. (See page 16)..	1.10	1.25	40	15	6
Citron. For preserves only ; hardy and very productive....				15	6
New Orange. Improved in size—the rind will peel like an orange when fully ripe...	1.10	1.25	40	15	6

MUSTARD.

Sow in drills one foot apart, and cover seed half inch deep. Thrives readily in almost any soil. Water frequently in dry weather, and for a succession sow every two weeks during the season. Used principally for salads.

Chinese. New. A fine sort...	1.10	1.25	40	20	10
White or Yellow. For salad or medicinal purposes...	45	60	20	10	6

NASTURTIUM.

Plant in May in rows; the climbing varieties to cover some arbor, or fence, or climb or twine around the house; the dwarf kind in hills or in rows two feet apart. The leaves are used for salad, and the seeds when soft enough to be easily penetrated by the nail, for pickles. If each plant of the dwarf variety is allowed room to perfect itself the plants grow very symetrical.

Tall. An ornamental climber...				15	6
Dwarf...				20	6

OKRA, or GUMBO.

Select warm and rich soil and plant when the ground becomes warm, in rows two feet apart, thinning plants a foot apart in the row. The pods are used to thicken soups, being gathered when young. In the North they require the warmest locations and it is better to start them in a hot-bed.

Early Dwarf. White, small and round ; pods smooth...	70	85	25	15	6
Long Green. Later and more productive...	70	85	25	15	6

ONION.

In some parts of the country the term "Silver Skin" denotes a white variety—in other parts a yellow variety. Please indicate in your order which you want. The soil should be rather light, and free from large stones. Apply from eight to twelve cords of rich, fine compost to the acre. Plough not over five inches deep, and work well with cultivator. Plough again at right angles with first furrows, and cultivate again. Now rake level and fine, and plant seed in rows fourteen inches apart at rate of four pounds to the acre. Keep very clean of weeds. When ripe, pull and dry very thoroughly before storing. For full particulars in every department for the cultivation of the Onion, see my work on "Onion Raising."

Southport White Globe. Remarkably handsome ; great cropper, but would not advise to raise it north of southern Connecticut. (See page 16)...	3.85	4.00	1.30	40	10
Marzajole. A silvery white skinned variety, possessing the best qualities desirable for culinary purposes...			1.25	40	10
Yellow Strasburgh, or Large Yellow. A late standard variety...	3.85	4.00	1.30	40	10
Mammoth Tripoli. One of the largest of the giant foreign varieties; of mild flavor...............package only				10	
Nasbey's Mammoth. An Italian variety excelling in size and mild flavor.................package only				10	
Giant Rocca. A new Italian variety. (See page 16)...				40	10
Large Flat White Italian. A mild flavored onion; grows from sets it attains to a very large size...				50	15
Early Red Globe. One of the earliest and most productive and handsomest of all the red sorts...	3.85	4.00	1.30	40	10
Early Cracker. A decided improvement on Large Yellow, being much earlier ; the kind for a short season...	3.85	4.00	1.30	40	10
White Portugal. Very early; mild flavored, not a good keeper...	3.85	4.00	1.30	40	10
Large Red Wethersfield. An old standard sort. Pleasant flavored, grows very large, keeps well ; hardy...	2.10	2.25	75	20	10
Early Flat Red. A capital sort where the seasons are short. A very quick grower...	3.85	4.00	1.30	40	10
Southport Red Globe. A great cropper. Very popular in New York market; is late would not advise to raise north of southern Connecticut...	3.85	4.00	1.30	40	10
New Queen. A new English white onion, the earliest of all varieties (See page 16)...	3.55	4.00	1.30	40	10
Danvers Yellow. (True.) Large, round, earlier than Large Yellow, very profitable; 1100 bushels have been raised from one acre...	4.85	5.00	1.50	50	10

Potato Onions. The bulbs of these are planted.........per peck, $1.50; per bush., $5.00; por qt., .25; qt. mail, .45.
Onion Sets. From these most of the early onions are raised......per qt., .50; qt., mail, .50; per bush., market price.

PARSLEY.

Select rich soil and sow the seed in drills one foot apart, covering half inch deep. The seed is usually from fifteen to twenty-five days in vegetating. Thin plants to four inches apart when two inches high. The beauty of the plant may be increased by several successive transplantings. It is used principally for flavoring soups, &c., and for garnishing in its natural state.

Fern Leaved. (New.) (See page 16.)...	1.60	1.75	60	20	10
Dwarf Curled. Finely curled ; good for edging or table ornament...	75	90	30	10	6
Myatt's Garnishing. Double curled...				15	6
Dunnett's Selected. A new English sort...				15	6
Carter's Champion. (New.) Moss curled elegant for garnishing...	1.35	1.50	50	25	6
Carter's Covent Garden Garnishing. Probably the best of its type...				25	6

PARSNIP.

Give the richest and deepest soil to the long varieties of Parsnip; the Turnip sort will grow well on shallow soil. Make the soil very fine, and plant the seed early in rows eighteen inches apart, thinning plants to five inches in the rows. The seed should be planted half inch deep. To keep well in the ground over winter, draw a little earth over the tops.

Sutton's Student. A good English variety...	60	75	25	12	6
Round Early, or Turnip. A new French sort ; excellent for shallow soil, shaped like a turnip...	85	1.00	30	15	6
Large Dutch. Large and sweet. A standard kind...	60	75	25	10	5
Hollow Crowned, or Guernsey. The hollow crowns are considered superior in quality to the other varieties...	60	75	25	10	6
Abbott's Improved Hollow Crowned. An improved English variety...	60	75	25	12	6
Maltese Parsnip. A new, long English variety...	60	75	25	10	6

PEAS.

Of the numbers printed against the Peas, 1 indicates first early class, 2 second early and 3 late class. Those marked with a star (*) are wrinkled varieties, the sweetest of all ; but as they are liable to rot need to be planted thicker than the round sorts. For notes on Peas see elsewhere

Very Dwarf. These very low varieties require no sticking.

While the tall sort will run too much to vine if liberally manured (it being better to depend on the richness of land that has been previously in good cultivation) the dwarf varieties, on the contrary, will bear pretty liberal manuring. Have the dwarfs, that grow not over fifteen inches high, in rows two feet apart ; those varieties attaining the height of from two to three feet, in rows three feet apart; and the rows of the tallest sorts, four feet apart.

	pack exp.	bush. exp.	qt exp.	qt. mail	pk'g
1 ***Carter's Extra Early Premium Pea.** A new early dwarf wrinkled pea, sent out by Messrs. Carter & Co., seedsmen, of London, as an improvement on Little Gem. More prolific and longer podded...	2.50	9.00	40	70	10
1. **Tom Thumb.** One of the very earliest ; very productive; pods well filled. Height of vine ten inches...	2.25	8.00	35	65	10
2. ***McLean's Little Gem.** A wrinkled pea nearly as early as Tom Thumb; quality first rate. Twelve inches...	2.25	8.00	35	65	10
1. **McLean's Blue Peter.** Early ; of fine quality ; pods larger than Tom Thumb, but not so numerous. Ten inches...			30	80	10

☞ Pounds, Bushels, &c., are priced on these pages instead of separate ones as formerly. 🖅

PRICES OF SEEDS.

PEAS.

Dwarf.

All varieties under this class will do without bushing, but on rich, garden soil they will generally do better when bushed.

	peck exp.	b'sh exp.	qt. exp.	qt. mail	pk'g
Dr. McLean's. (New.) See page 15..			50	80	15
3. Carter's Challenger. (New.) (See page 17.)................................package only					20
2. Carter's Little Wonder. (New.) (See page 17.)..			50	80	20
1. Hancock. This is a new seedling of American origin. A first early, and, all things considered, the best of the early hard peas. See notes elsewhere................................	2.12	7.50	35	65	10
2. Fill-Basket. (New.) A large, very handsome and productive sort ; very prolific............			43	75	10
1. Sutton's Emerald Gem. (New.) (For description see page 17.)................................			45	75	10
1. Philadelphia Extra Early. The standard early variety in Philadelphia markets............	2.00	7.00	30	60	10
1. Carter's First Crop. Earliest of all; pods smaller and more numerous than Dan O'Rourke. Two and one-half feet.	2.00	7.00	35	65	10
1. Extra Early Dan O'Rourke. One of the earliest standard market varieties ; very productive. Two feet.	2.00	7.00	30	60	10
1. Kentish Invicta. A new English variety, very early and of great promise. Crop ripens all together. Two and one-half feet.	2.25	8.00	35	65	10
1. *Laxton's Alpha. The best early wrinkled market pea. In yield it probably surpasses any of the early sorts........	2.25	8.00	40	70	10
1. Caractacus. Messrs. Waite & Co., the English seedsmen, send this out. It is planted largely by the Boston market-men as one of the best first early peas. Two feet.	2.25	8.00	35	65	10
1. Dexter. A new American pea, selected as being extra early. Worthy of a trial by gardeners. Two and one-half feet...	2.25	8.00	35	65	10
2. *McLean's Advancer. A wrinkled pea—about a fortnight earlier than Champion of England, equal to it in quality, fully as productive, while it grows but two-thirds as high ; everything considered, the best of the second earlies for market purposes. Two and one-half feet. Very popular both for the family garden and for market.	2.25	8.00	40	70	10
2. *Hair's Dwarf Mammoth. One of the best for family use—low and bushy in its habit of growth ; peas very large, wrinkled and sweet. Eighteen inches.	2.25	8.00	40	70	10
2. Brown's Dwarf Marrowfat. The earliest of all marrowfats ; dwarfish habit. A first class American variety. Two feet.	2.25	8.50	45	75	10
3. *Yorkshire Hero. A large late wrinkled dwarf; peas remarkably large and fine ; a capital sort for the kitchen garden. Two and one-half feet	2.25	8.00	40	70	10
3. *McLean's Premier. An English wrinkled pea, pods and peas very large; sent out as being of very superior quality and productiveness. A nice family pea. Two and one-half feet.	2.50	9.00	45	75	10
2. Dwarf Blue Imperial. An old standard sort ; two feet.			30	60	10

Tall Varieties. All these need bushing.

3. *Carter's Commander-in-Chief. New. (See page 17.)..............................package only					15
3. Culverwell's Telegraph. (New.) See page 16. Pods are exceptionally large and well-filled....package only					15
3. New Golden. New. (See page 17.)..			40	70	15
3. Laxton's Superlative. (New.) (For description see page 16.)................................	2.50		45	75	15
3. Dwarf Sugar. A string pea; pods edible. My variety is of half dwarfish habit, with fine large pods............			45	75	15
3. Laxton's Supreme. One of the green marrow class yielding remarkably long and well-filled pods. A fine late family garden pea. Five feet.			45	75	15
3. *Champion of England. An old favorite; rich flavored and very productive. Four to five feet............	2.00	6.00	30	60	10
3. Black Eyed Marrowfat. An old favorite ; large podded ; prolific ; capital for market. Three to four feet........	75	2.50	20	50	10
2. Royal Dwarf Marrowfat. Not so tall as Large White Marrowfat; earlier than Champion of England........	1.30	4.00	23	55	10
3. Large White Marrowfat. A standard late sort.	1.00	3.00	25	55	10

PEPPER.

Peppers should be started in a cold frame or hot-bed. Transplant the young plants into the open ground towards the close of May in a very warm location, having the rows eighteen inches apart; thin plants a foot apart in the rows. The ground should be made very rich, either by high manuring before plants are transplanted, or by liberal application of guano, liquid manures afterward.

	lb. exp.	lb. mail	¼lb.	oz.	
Spanish Monstrous. On good soil will grow six inches long and two inches in diameter. See page 13............			40	15	
Chili. Sharply conical, about two inches in length and one-half inch in diameter. Of a brilliant scarlet color when ripe.			40	6	
Long Yellow.			40	6	
Large Bell. A standard sort.	3.35	3.50	1.00	35	6
Cayenne. Small, long and tapering ; very hot ; best for seasoning pickles.	3.35	3.50	1.00	35	6
Large Sweet Mountain. Very large and excellent for mangoes.	3.65	4.00	1.30	40	6
Cherry. Small, smooth and round ; a great bearer.			35	6	
Squash, or Flat. The variety generally planted for family use ; large and thick fleshed ; the best for pickling...	3.35	3.50	1.00	35	6
Long Red, or Sante Fe.			35	10	

PUMPKIN.

Cultivate as Squash, which see for general directions.

Negro. (New) (For description see page 17.)..	1.35	1.50	40	15	6
Tree. (New) (For description see page 17.)..			30	10	
Large Field. Good for stock. per quart, 50 cents..........	25	40	15		6
Sugar Pumpkin. Smaller than Large Field, but fine grained, sweeter and very prolific; first rate either for the table or stock.	85	1.00	30	15	6
Cheese. A variety popular in the Middle States. Cheese-shaped, resembling in character the Crookneck Squash...!	60	75	25	10	6
Michigan Mammoth. A soft shelled variety, excellent for stock It grows very large and is a heavy cropper........				35	10

RADISH.

For early use sow in spring, as soon as the ground can be worked, in drills six to ten inches apart, covering seed half inch deep. Thin plants an inch apart in the row. As the roots are more succulent and tender when grown quickly, a rich, light soil should be preferred and frequent watering in dry weather will be found beneficial. For a succession sow every two weeks. The Olive shaped varieties are more tender, sweeter and earlier than the long kinds, and not so apt to be worm eaten.

White Russian Winter. New. See engraving on page 11..	2.35	2.50	75	25	10
Carter's Selected Long Scarlet. Sent out by Messrs. Carter & Co. as the best variety of Long Scarlet. Has proved a favorite among market gardeners around Boston.	85	1.00	30	12	6
French Breakfast. A beautiful variety of the Olive radish, scarlet in the body and white at the extremity........	85	1.00	30	12	6
Wood's Fine Frame. Excellent for cultivation under glass ; very early. A favorite with English market gardeners. In shape between Olive and Long.	65	80	25	12	6
Covent Garden. A fine selection of Long Scarlet. This new sort is considered the best of all the Long Scarlet varieties.	60	75	25	12	6
London Particular Long Scarlet. Held in high esteem in London market ; fine, long scarlet............	85	1.00	30	12	6
Early Scarlet Olive Shaped. Very early and handsome ; quick growth, tender, excellent. A favorite........	85	1.00	30	12	6
Golden Yellow Summer Turnip. New. See page 16. Color very rich; very early............	1.10	1.25	35	15	6
Early Rose Olive. Differs from Early Scarlet Olive in color only....................................	85	1.00	30	12	6
Red Turnip Rooted. A standard early, very popular in markets of New York.	85	1.00	30	12	6
White Turnip Rooted. For summer and winter use.	60	75	30	12	6
Black Spanish. Round variety.	60	75	25	12	6

☞ Pounds, Bushels, &c., are priced on these pages instead of separate ones as formerly. ☜ **PRICES OF SEEDS.**

RADISH.

	lb. exp. mail	lb.	¼ lb.	oz.	pk'g
Yellow Summer Turnip. An early and excellent summer variety.	.85	1.00	30	12	6
Chinese Rose Winter. The best for winter use. Grows large and tender.	1.85	1.50	50	20	10
Raphanus Caudatus, or Rat-tailed Radish. Pods grow to a foot or more in length, and are edible, package only					15
California Mammoth White. (New.) A new winter sort, eight to twelve inches long, and two inches in diameter in the largest part. From the Chinese in California. An acquisition.	1.60	1.75	60	20	10

RHUBARB.

Sow the seed in drills eighteen inches apart and cover one inch deep. Thin the plants to a foot apart. When the plants are one year old prepare the ground for the final bed by trenching two feet deep, mixing a liberal quantity of manure with the soil. Set plants five feet apart each way. Do not cut until the second year, and give a dressing of manure every fall. If it is desired at any time to increase the bed, the roots may be taken up in the spring and divided. The seed will not always give plants like the parent.

Linnæus. Large, tender, and of excellent flavor. A well-known market variety.				30	15
Mammoth. The largest of all.				50	15

SALSIFY, or VEGETABLE OYSTER.

	1.60	1.75	60	20	10

Sow in early spring on light, rich soil, in drills fourteen inches apart and thin the plants to three inches in the row. The roots will be ready for use in October and will sustain no injury by being left in the ground during the winter. When cooked the flavor somewhat resembles the oyster.

SORREL. Large Leaved French.

				15	6

Sow in hot-bed early in the spring, and transplant to the open ground, on warm, mellow soil, when the ground has become warm, setting the plants in rows two feet apart and about sixteen inches apart in the row. As the seed is [rather slow to germinate, it should be watered liberally in the hot-bed.

SPINACH.

For summer use sow early in spring, in drills eight inches to one foot apart, covering the seed one inch deep. Select rich soil, and manure liberally. A succession may be obtained by sowing at intervals of two weeks through the season. For very early spring use sow in August. The plants are sometimes protected through the winter by a thick covering of straw or some similar, light covering. Spinach is used principally as greens for boiling, and is very highly esteemed for this purpose.

New Zealand. Makes a very large plant and will endure drought ; best quality. By some this is thought to promise well as a forage plant.	.85	1.00	30	15	6
Prickly Seeded. The hardiest variety ; thick leaved—for fall sowing.	.45	.60	20	10	6
Round Leaved. The popular summer variety.	.55	.50	20	10	6
Extra Large Round Leaved. (New.)	.50	.65	20	12	6

SQUASH.

All vines delight in warm and rich soil. Prepare the ground by thoroughly pulverizing. Manure at rate of six or eight cords to the acre, working it just under the surface with the cultivator or gang plough. Plant in hills nine to ten feet apart for running varieties, and five or six feet apart for bush sorts; work some rich, fine manure into each hill. Leave two plants to the hill. Keep well covered with plaster or air-slaked lime in early stages of growth. Cultivate frequently until runners are well started. For full particulars in every department, see my work "Squashes and How to grow them."

Essex Hybrid. New. See page 14.	2.85	3.00	1.00	40	15
White Early Bush. The earliest sort.	.80	.95	30	12	6
Summer Crookneck. Early, fine for summer use.	.80	.95	30	12	6
Golden Bush. A fine early summer sort.	.80	.95	30	12	6
Vegetable Marrow. The standard English squash ; a fair summer variety with us ; a heavy cropper ; good for stock.	1.35	1.70	40	15	6
Cambridge Marrow. (New. See page 17.)	1.15	1.70	40	15	6
Boston Marrow. A standard fall squash ; of a rich orange color, and very productive.	1.15	1.30	40	15	6
× **American Turban.** Decidedly the best of all fall squashes. (See page 12.)	1.15	1.30	40	20	10
The Butman. (New. For description see page 6.).	1.35	1.50	50	20	10
Marblehead (New. For full description see page 7.).	1.15	1.30	40	20	10
× **Hubbard.** A standard winter squash. (See page 12.)	1.35	1.50	50	20	6
Cocoanut. (See page 7.) A half-bush variety of small size, but remarkably heavy and very prolific ; fine grained and of a very rich, chestnut-like flavor This squash is worth raising as an ornament for the parlor.	2.35	2.50	1.00	30	10
Yokohama. This squash has the flavor of the crookneck class, but is finer grained and much superior in quality.				30	10
Mammoth Yellow. Has been grown to weigh from 100 to 300 pounds	3.85	4.00	1.30	40	15
Canada Crookneck. The small, well-known, excellent kind.				30	10
Large Winter Crookneck. The old standard sort, the best of keepers ; cross grained.	.85	1.00	30	15	6

SUNFLOWER.

Sow thinly in drills three feet apart and thin two to three feet apart in the row.

Common.	.85	1.00	30	10	6
Mammoth Russian. A fine variety. Valuable for poultry or vegetable oil. Flowers and seed very large.	1.35	1.50	50	20	10

SWISS CHARD.

Plant and cultivate as Beets. The tops while young are boiled as greens and the center leaf cooked and served like asparagus.

Ornamental Varieties. The leaf veins are white or brilliant scarlet and yellow, and are planted in England scattered through flower plots with fine effect. package only					10
Common Chard				15	10

TOMATO. ☞ All of the varieties of Tomato seeds are of my own growing from carefully selected stock.

Sow the seeds in March or April in the hot-bed or in pots in a sunny exposure in the house. If it is desired to make very healthy, stocky plants, they may be transplanted or repotted when about two or three inches high. When five or six inches high, if the ground has become warm, transplant to the open ground, on a rainy or cloudy day if possible, if not, the young plants should be liberally watered and shaded from the hot sun. If it is desirable to have the fruit ripen as early as possible, in preference to a heavy crop, select rather light, poor soil and a sunny location. Trimming off the laterals, training to a stake and propagating by slips, are believed by many to promote earliness. Set the plants four feet apart each way, upon mounds of earth, to allow the foliage to open and let the sun in amongst the fruit. A cheap trellis made by driving three stakes around the plants and encircling them with three or four barrel hoops makes a very nice support for training them on.

Alpha. New. See engraving elsewhere. Every gardener should have it.				1.50	20
Alpha. Extra selected seed from very earliest fruit.				2.00	20
Red Chief. New. See page 15.					15
Triumph. New. See page 15. package only					15
New Japanese. New. See page 13. I find nothing specially desirable in this ; perhaps others may have sharper eyes.					10
Criterion. New. Closely allied to the peach variety.					10
Yellow Victor. New. (See page 17.)				40	10
Little Gem. (New.) A small variety, claimed to be extra early.	2.85	3.00	1.00	30	10
· **Powells.** New. (See page 17.)				50	10
Foote's Hundred Days. (New.) Fruit small and irregular ; remarkable for its earliness.				50	10
Conqueror. New. (See page 17.)	2.85	3.00	1.00	30	10
Livingston's Acme. (New) (See page 17.)	3.85	4.00	1.30	50	15

☞ Seeds per Express or Freight at purchaser's expense. 🖅 | PRICES OF SEEDS.

TOMATO.

	lb. exp. mail	lb. mail	¼lb.	oz.	pk'g
Paragon. (New. For description see page 7.)	3.85	4.00	1.30	50	10
Canada Victor. (New.) For full description of this fine, new tomato see page 12.)	4.35	4.50	1.25	45	10
Canada Victor. Extra selected seed. (See page 12.)......per lb. by express, $11.85.	$12	4.50	1.25		15
Hathaway's Excelsior. One of the spherical, round tomatoes, being quite early, uniformly round, very solid, of excellent quality and very productive. Skin rather thin	3.35	3.50	1.00	35	10
Arlington. (New.) Of large size, quite smooth and solid; late				40	10
Essex Early Round. Very early, round and solid. A capital sort for early market; very popular in eastern Massachusetts				40	10
Hubbard's Curled Leaf. This tomato so closely resembles Early York that I consider it but a strain of that variety.				40	10
Trophy. This magnificent variety is particularly valuable in the home garden.	4.35	4.50	1.25	45	10
Gen. Grant. Smooth, symmetrical and solid. A popular market sort around Boston	2.85	3.00	1.00	40	10
Orangefield. A new English variety of a rich fruity flavor.				40	10
Early York. Very early, dwarf and productive: somewhat irregular in shape				40	10
Keyes' Early Prolific. One of the earliest; bears its fruit in large clusters of medium sized tomatoes.	2.85	3.00	1.00	40	10
Yellow Fig. Pear shaped, and used to preserve as figs					10
Cherry. Flavor unsurpassed. Fruit small, but a wonderful cropper......package only					10
Mammoth Chihuahua. Grows to weigh as high as two pounds each. More curious than useful......package only					5
Cook's Favorite. Roundish in shape and quite solid when fully ripe.				40	5
Boston Market. I now have a first rate strain of this variety; early, large, smooth and solid.	2.85	3.00	1.00	30	10
New White Apple. Sweet, with a rich, fruit-like flavor. Highly ornamental for the table......package only					5
Large Yellow. Large and of a sweetish and peculiar flavor.					5
Large Smooth Red. The standard kind; good for market purposes	2.85	3.00	1.00	30	5
Tilden. First class on low, rich soil; of large size, thick meated, smooth and of a high flavor.	3.85	4.00	1.30	40	5
New Mexican. Large, round; a good variety for the South as the fruit is protected by the leaves.				45	5
Strawberry, or Ground Cherry. (Alkekengi.) Grows enclosed in a husk; excellent for preserves; will keep within husks all winter					10
Yellow Plum. Small, elegant. Nice for preserve......package only					10
Currant. Very elegant; resembles long bunches of currants; for ornament only......package only					10

TURNIP.

For early use sow the small sorts as soon as the ground can be worked in the spring, in drills fourteen inches apart—the Rutabagas thirty inches. As the seed is very fine it should be covered but slightly, excepting in very dry weather. Select light. If possible, new soil, and manure with plaster and ashes, or phosphates. Should the young plants be troubled with insects, a sprinkling of the same will be found beneficial. Of the early varieties thin the plants to six inches apart and the Rutabaga to one foot. For fall and winter use the early kinds should be sown from the middle of July to the middle of August, and the Rutabagas from the middle of June to the first of July, using from one to one and a half pounds of seed to the acre. Turnips are very extensively used as winter feed for cattle and sheep. "Swede" and "Rutabaga" being synonymous terms, the names below, given as they are generally used, will be readily understood. The English varieties are almost exclusively used for early planting for market.

English Varieties.

	lb. exp. mail	lb. mail	¼lb.	oz.	pk'g
White Egg. (New.) (See page 5.) Large, handsome, early; keeps first-rate......per half pound, 60 cents.	85	1.00	40	15	10
Carter's Stone or Stubble. Almost as early as White Top or White Dutch, but much thicker; handsome.	60	75	25	10	6
Golden Stone. A fine, oblong, yellow-fleshed variety; very handsome; an acquisition	60	75	25	10	6
Pomeranian White Globe. A very fine white globe variety; round and thick	60	75	25	10	6
Early White Dutch, or Early Flat Dutch. Allied to the White Top; of medium size and quick growth.	60	75	25	10	6
Early Red Top. Fine, sweet, mild, rapid grower; very early and popular.	60	75	25	10	6
Early White Top. Differs from Red Top only in color	60	75	25	10	6
Improved Yellow Globe. Fine for family use or for field culture. An excellent American variety.	75	90	30	10	6
Orange Jelly. A round, yellow English turnip of finer quality than Golden Ball	60	75	25	10	6
Yellow Finland. Very elegant; less eaten by worms than most sorts.				12	6
Long White Cowhorn. Matures quickly, carrot shaped, fine grained and sweet	60	75	25	10	6
White Tankard. One of the English varieties—in shape resembling a tankard. White-fleshed; a very heavy cropper.	60	75	25	10	6
Jersey Navet. A new English variety of white turnip; an underground turnip; early, very sweet.	85	1.00	50	10	6
Green Top Aberdeen. Round, yellow-fleshed and firm; a good keeper. This variety in earliness and for stock purposes is half way between the early varieties and the Rutabaga. It does not succeed in all localities.	60	75	25	10	6

Swedes, or Rutabagas.

	lb. exp. mail	lb. mail	¼lb.	oz.	pk'g
Hartley's Swede. One of the largest of Swedes; short-necked; round in shape; very fine; an acquisition	70	85	25	12	6
Golden Swede. Has a small top, fine root and thin rind and ranks high in England.	70	85	25	12	6
American Rutabaga. Popular among our market gardeners both for table and for feeding stock. Flesh very solid. A first-rate keeper.	60	75	25	10	6
Laing's Improved Swede. One of the earliest swedes; a handsome variety of excellent quality. Yellow fleshed. For the table.	60	75	25	10	6
Large White French or White Swede. A white market rutabaga.	60	75	25	10	6
Sweet German or Russian. White, sweet, excellent—a first rate keeper.	60	75	25	10	6
Skirving's Purple Top Rutabaga. A standard field variety for stock and fine for family use. Yellow fleshed.	60	75	25	10	6
Shamrock Swede. A popular English sort. Short neck and oblong in shape. Yellow fleshed.	60	75	25	10	6
London Purple Top Swede. Short neck and round in shape; grows to a larger size than Skirving's and has a shorter neck. Yellow fleshed.	65	80	25	10	6
Carter's Imperial Swede. Messrs. Carter & Co. rank this as the best of their swede turnips.	65	80	25	10	6

VEGETABLE SNAILS AND CATERPILLARS......package only | | | | | 10

WATER CHESTNUT. (Trapa Natans.) New. To be grown in the mud of a brook; edible. See page 15. package only | | | | | 10

WHITLOOF. New. (See page 17.)......package only | | | | | 10

SWEET OR POT AND MEDICINAL HERBS.

	lb. exp. mail	lb. mail	¼lb.	oz.	pk'g		oz.	pk'g		pk'g
Sage—American seed	2.35	2.50	75	25	6	Coriander	10	6	Belladonna	6
Thyme, broad leaved English	3.85	4.00	1.30	40	10	Sweet Basil	15	6	Cumin	6
Summer Savory	1.85	2.00	70	30		Dill	10	6	Fenugreek	6
Sweet Marjoram	2.85	3.00	1.00	30	6	Hyssop	30	6	Henbane	6
Caraway	85	1.00	30	10	6	Rue	50	10	Melis Balm	10
Hoarhound	3.85	4.00	1.30	40	10	Lavender	30	5	Wormwood	10
Saffron	1.35	1.50	40	15	6	Rosemary	50	10	Fuller's Teasel	1
White or Opium Poppy	1.85	2.00	70	20	6	Sweet Fennel	10	6		

☞ Pounds, Bushels, &c., are priced on these pages instead of separate ones as formerly. ☜

PRICES OF SEEDS.

GRASS, AND CLOVER SEEDS, &c.

	peck/b'sh	3 lbs.	1 lb.	pk'z
	exp.	exp.	mail	mail
White Zealand Oats. New. See third page of cover	.50	1.75	.90	.35	10
Hulless Barley. New. See third page of cover	3.00	$10	1.00	.30	15
Defiance Wheat. New. See page 14	3.00	$10	1.00	.50	10
Champlain Wheat. New. See page 14	3.00	$10	1.00	.50	10
White Russian Spring Wheat. (New.) (See page 5.) 10 bush., $2.25 per bush., per 1-2 bush., $1.50	.80	2.65	.90	.35	10
Chinese Hulless Oats. (New.) (See page 16.)	.75	2.25	.90	.35	10
Probsteier Oats. (See third page of cover.)	.50	1.25	.90	.35	10
Alsike Clover	4.00	$15	1.25	.50	10
Red Clover			.90	.35	10
White Clover			1.75	.75	10
Lucerne. (See page 15.)			1.50	.70	10
Silver Hull Buckwheat. (New.) Husks thinner than those of the common sort. Stands drouth first-rate	.70	2.80	.90	.35	10
Hungarian Grass. A magnificent forage crop for light land. On land in high condition, two crops may be grown the same season.	.70	2.80	.90	.35	10
Pearl Millet. See page 16	3.00	$10	1.00	.30	10
German or Golden Millet. A heavier cropper than Hungarian, leaves broader and stalks stouter; it requires strong land, whereas Hungarian will do well on light land	.80	2.50	.90	.35	10
Timothy Grass			.90	.35	10
Red Top Grass			.90	.35	10
Orchard Grass			....Quart per express, 25; per mail 35.	1.00	3.00		10
Lawn Grass. Flint's mixture of fine grasses. From Grasses and Forage Plants. By Hon. C. L. Flint, Secretary Mass. State Board of Agriculture	....Quart per express, 40; per mail 50.	1.75	6.50		10
Lawn Grass. Common Mixture	....Quart per express, 20; per mail 30.	1.25	4.00		10
Giant Summer Rye. Kernels enormously large; a Spring rye			.90	.35	10
Rape Seed			.90	.35	10
Cranberry Vines. See third page of cover					
Grafting Wax			.55		

☞ Prices for Red Clover, Lucerne, Timothy and Red Top in quantity given upon application. ☜

HEDGE AND TREE SEEDS.

Keep seed of Evergreen in dry sand until time of sowing. Sow these early in spring in well-prepared beds of sandy loam, covering to the depth of about the thickness of the seed, pressing the earth firmly over them. Water frequently but not much at a time. Keep down all grass and weeds, and the first season protect with brush or lathe screens from the sun's rays. Transplant into nursery rows when two years old. Plant seed of deciduous trees early in spring in drills about two feet apart. The young plants will not require shading. Acorns, Chestnuts and Walnuts should be planted in autumn, or be kept from shrivelling up over winter in sand or moss. Seeds of Locust, if not planted in autumn, require the action of frost on them. Seed of the American Elm may be planted as soon as they fall from the tree. The Berberry vegetates best when planted in autumn.

	lb. exp.	lb. mail	1 lb. oz.	oz. pk'z		oz. pk'z		oz. pk'z
Apple	1.35	1.50	50	15	6	American Arbor Vitæ	50	10	Magnolia	40	10
Pear	2.85	3.00	1.00	30	10	White Ash	20	6	Norway Maple	15	6
Berberry. See page 20	1.10	1.25	40	15	6	American Beech	10	6	White Oak		15
Honey Locust	.85	1.00	30	10	6	White Birch	40	10	White Pine	60	10
Yellow Locust	.85	1.00	30	10	6	American Elm	50	10	Pitch Pine		10
Osage Orange	.60	.75	20	10	6	Eucalyptus Globulus		10	Scotch Pine	30	6
Sugar Maple	1.10	1.25	35	10	6	Balsam Fir	25	6	Hemlock Spruce	50	10
Red Cedar	.60	.75	25	10	6	Tree of Heaven	40	10	California Big Tree		25
Shellbark Hickory	.15	.30				European Larch	25	6	Tulip Tree	30	10

ASPARAGUS AND HORSE-RADISH ROOTS.

Defiance two years old. See page 14. Price per 100 $3.00.
Conover's one year old. Price per 100, $1.25; per 1,000, $10.00; small roots, per 100 per mail, $1.60.
Conover's two years old. Price per 100, $1.50; per 1,000, $12.00; the two years roots are too bulky to send by mail.
Horse-radish. Per 100, $1.00; per 1,000, $7.00; per 100, per mail, $1.50.

STRAWBERRY VINES.

Select rather moist soil, dig deep, make fine, manure with rotted manure, bone or wood ashes. For hill culture plant 15 by 15 inches, and pinch off the runners as fast as they appear; for matted growth, plant rows three feet apart, and plants one foot apart in the row, and let runners grow. The hill makes the largest berries, the matted system gives the greatest crop, and is best for light soil.

	PRICES.		
	1,000 exp.	100 exp.	100 mail
CRESCENT SEEDLING. New. Wonderfully productive, equaling or surpassing Wilson. Size above medium		2.00	2.40
SHARPLESS SEEDLING. New. Highly praised; a very vigorous, healthy grower; berries large and of good quality		2.00	2.40
GREAT AMERICAN. (New.) The berries of this variety probably excel in size any of late introduction	$12	1.50	1.90
BELLE. Fruit large; smooth and handsome; a good bearer	8.00	1.00	1.40
CAROLINE. Medium early; fruit large, smooth and sweet; handles well; Very prolific; compact habit of vine	8.05	1.00	1.40
WILSON'S ALBANY. The great market berry; does well everywhere	8.00	1.00	1.40
CHARLES DOWNING. Vigorous, hardy and productive; fruit uniformly large, juicy and of excellent flavor	8.00	1.00	1.40
JUCUNDA. Fruit very large and heavy; often commands highest price in the market. Requires rich soil and high cultivation	8.00	1.00	1.40
BLACK DEFIANCE. Very early; large, deep scarlet fruit	8.00	1.00	1.40
GREEN PROLIFIC. A good variety for light soils; very hardy; fruit very sweet	8.00	1.00	1.40

PRICE LIST OF POTATOES. See Description Elsewhere.

	bbl. exp.	b'sh exp.	peck exp.	25 eyes mail	1 lb. mail	3 lbs. mail		bbl. exp.	b'sh exp.	peck exp.	25 eyes mail	1 lb. mail	3 lbs. mail
Clark's No. 1. New	5.00	2.50	1.00	50	50	1.00	Beauty of Hebron. New	4.00	2.00	75	50	50	1.00
Late Ohio. New	4.50	2.25	80	50	50	1.00	Early Ohio	3.75	1.84	75	50	50	1.00
Moore's Seedling. New	5.00	2.50	1.00	50	50	1.00	Burbank's Seedling	3.75	1.88	75	50	50	1.00
Clark's No. 2. New	4.50	2.25	75	50	50	1.00	Dunmore Seedling	3.75	1.88	75	50	50	1.00
Mammoth Pearl. New	4.50	2.25	75	50	50	1.00	Extra Early Vermont	3.25	1.75	75	50	50	1.00
White Rose. New	4.00	2.00	75	50	50	1.00	Bresee's No. 6, or Peerless	3.25	1.75	75	50	50	1.00
Bliss' Triumph. New	5.00	2.50	1.00	50	50	1.00	Early Rose	3.00	1.50	75	50	50	1.00

SUPERIOR ONION SEED.

Sow in early Spring in drills 14 inches apart, using 4 lbs. of seed to the acre.

My stock of Onion Seed is raised with peculiar care, none but the very best and earliest onions being selected for seed stock, and from these I select carefully, by hand, my seed onions. I have the best grounds for believing that *not an ounce of seed is sent out from my establishment that is not perfectly reliable*; and second, exercising such care, I claim for it a quality superior to most of that in the market, which is raised either from very small, refuse onions, from large and coarse onions, or, again, from such onions as the seed grower chances to have on hand—either of which qualities tends to produce poor onions. All the onion seed of my own growing is raised in locations perfectly isolated, and the *yellow sorts will be found to be almost entirely free of any mixture with red*. There is a good deal of cheap seed in the country again this season, there having been a great quantity of old seed carried over from last season. Such seed experienced gardeners will not purchase *at any price*. Below I add a few extracts from the many letters received from my customers.

Early Round Danvers Yellow Onion. The Danvers Onion excels in earliness, and yields more than the Flat Red or Yellow sorts. Over 1000 bushels have been raised on an acre. It commands in Boston market a readier sale than the Red sorts, and brings a higher price in the market.

Early Flat Red. One of the very earliest, hardy and reliable. A capital sort for the extreme north where other sorts are too late to bottom and ripen well.

Early Red Globe, or Danvers Red. This ripens two or three weeks earlier than Red Wethersfield. Scullions are almost unknown with this onion when grown from most carefully selected seed stock, it being as reliable for bottoming well as Winnigstadt Cabbage to make a head.

Southport Late Red Globe. This variety of Late Red is very popular in the markets of New York, where it sells at a higher price per barrel than the common Red Wethersfield. Being much rounder it measures better, and for this reason also is more profitable for market gardeners to grow. It is quite late, and I therefore do not recommend it for farmers who live north of southern Connecticut.

White Globe. As white and handsome as a newly made snow-ball, sells half as high again as other sorts. Too late to grow north of southern Connecticut. Should be dried in an airy and shady place.

The Early Cracker Onion Is the earliest of all my Yellow sorts and an excellent kind to raise where the seasons are short.

PEDIGREE ONION SEED.

An observing intelligent farmer claimed that earliness, reliability for bottoming, smallness of neck can be as thoroughly inbred in an onion, as capacity to transmit her good qualities can be inbred in a cow or in any class of animals. What is possible to attain to in this matter my customers find in my onion seed, as shown in *the extracts from letters given below*. Those who have never raised onions from seed raised from stock so carefully selected will be equally surprised and pleased at the result.

An axiom that every onion grower soon learns, "*cheap onion seed is always dear.*"

The stock of Onion Seed is quite short this season, and this is especially true of the Early Round Yellow Danvers. My crop is smaller than usual, and I would therefore advise my customers to order their supply at an early day.

ONION SEED BY THE POUND.

	¼ lb.	lb.	Mail.	Exp.
White Globe	$1 30	$4 00	$3 85	
New Queen	1 30	4 00	3 85	
White Portugal	1 50	4 00	3 85	
Large Red Wethersfield (own growing)	75	2 25	2 10	
Southport Late Red Globe	1 30	4 00	3 85	
Early Round Yellow Danvers, my own growing, from hand picked onions	1 50	5 00	4 85	
Extra Early Yellow Cracker, my own growing..	1 30	4 00	3 85	
Early Red Globe, (own growing)	1 30	4 00	3 85	
Early Flat Red, my own growing	1 30	4 00	3 85	

For 5 lb. lots of the above **except Early Round Yellow Danvers, my price will be 25 cts. less per lb.**

THE SOUTH CAN RAISE ONIONS *the same year* of planting from the black seed. Please observe that several of the recommendations given below are from Southern States; and the onions were matured *full sized the first year* from my black seed. My customers in Texas have written me that they have grown onions the first year from black seed of my raising, in one instance, three inches in diameter, and in another weighing a pound each.

"I Raised Yellow Cracker Onions, 4½ inches in diameter, from seed bought of you." HENRY R. DEXTER.
Palestine, Anderson Co., Texas.

"I raised Onions, from seed bought of you last year, measuring 4½ inches in diameter, from black seed of the Early Red Globe Variety." H. B. SNYDER.
East Liberty, Pa.

"The seeds I had of you last year gave good satisfaction. From one ounce of Early Flat Red Onion seed I raised eleven bushels." D. G. BARTLETT.
Pretty Marsh, Hancock Co., Maine.

"The seeds I have purchased of you for three years have given satisfaction in every respect. I gathered over four hundred bushels per acre of Danvers Yellow Onions and of Early Flat Red. A. BRADLEY.
Langoase, Texas.

"The Onion seed were superior to any I ever had. Where in the garden is economy so misdirected as in buying cheap onion seed?" S. W. SEAMAN.
Mott's Corners, N. Y.

"Please send me two pounds of Early Round Yellow Danvers Onion seed. I can get seed in my own neighborhood for two dollars per pound, but prefer yours at four dollars. The seed I got of you last year did well. From three pounds of seed I grew 430 bushels." N. P. WATTS.
Perry, Ohio.

"Your Onion seed I think is superior to any in the market. My Yellow Danvers and Round Red which I raised last year from seed purchased of you, was awarded the first premium at the County Fair, last fall." J. B. STETSON.
Hanley, Minn.

"The Early Cracker Onion seed I had of you last year did nicely for this part of the State. I raised some that were six inches in diameter, and I think they are just the sort to raise where the season is as short as it is here." JAMES A. DODGE.
Sedgwick, Me.

"The seed I purchased from you last spring gave entire satisfaction, although the season was a very bad one. I think I will make 200 bushels of nice full grown onions from the four pounds of Early Cracker seed you sent me." JOHN F. WILLIAMS.
Bonsacks, Va.

"I have had extra good luck with all the seeds I have received from you. Onions do remarkably well, your seed being the only seed from which we can raise good onions the first year." J. M. L. PARKER.
Ahnepee, Wis.

"I have raised here from your Danvers Onion seed bulbs averaging three inches in diameter, the first year from seed. Had as good success as I ever had in the best onion growing section of Vermont." FRANK M. GAGE.
Rural Dover, Greensville Co., Va.

"I have bought Danvers Yellow Onion seed from you the last three years, and it has proved true to name every time. I could have bought onion seed last year for from one to two dollars per pound, but I was afraid of it. My experience is, good seed true to name is cheap at almost any price." HENRY SIVERS.
Oswego, N. Y.

"I must say that all the seeds purchased of you did extra well, especially the Danvers Onion, which grew larger from the seed than any onion around here grew from the sets." JAMES McGOWAN.
Phillipsburg, Warren Co., N. J.

"From your Danvers Onion seed planted in February of last year, I grew specimens weighing over one pound each." J. W. FLEMING.
San Saba, Texas.

"Your Danvers Onions grew better than those from the seed I raised on my own farm. It was the largest crop of onions I ever saw grow. I measured off three square rods of them in which there was not much difference; one of the pieces had eight bushels, wanting ten pounds, which is ahead of anything I ever saw in line of onions." I. F. JOSLIN.
Frankfort, New York.

Implements for Cultivation of Onions and other Vegetables.

MATTHEWS' SEED DRILL, No. 1., ...	$13 00
RUHLMAN'S IMPROVED WHEEL HOE........................	5 50
SCUFFLE HOE..............	1 00
DOUBLE WHEEL HOE......	1 70

The prices of these implements are those at which they are retailed at the seed stores in the large cities, at which rates I will superintend without charge the purchasing and forwarding of any of these machines to the address of any person ordering them.

Matthews' Garden Seed Drill.

The new drill was invented by Mr. E. G. MATTHEWS, the inventor and manufacturer of the HOLBROOK "REGULATOR" SEED DRILL, and has been remodelled and improved for 1877. Time and experience have enabled him to improve upon that and produce in this new machine one, which though closely resembling that, *is greatly superior to it*, or any other drill I am acquainted with. It is tasty in appearance; simple in principle; may be operated with ease and rapidity; is thoroughly constructed throughout, and there are no parts subject to unusual wear.

The Agitator is an ingenious and effective device, without springs, cams, gears or belts to get out of order, wherein it surpasses the Holbrook machine.

A simple contrivance gauges the deposit of the seed with mechanical accuracy to the depth required. Its Adjustable Markers answer the double purpose of markers and legs.

Full directions for use on the cover of each machine. Price, boxed and delivered at cars or boat, $13 00.

Ruhlman's Improved Wheel Hoe.

This superior implement embraces the four essential points requisite for a successful Hand Cultivator, viz.: Durability, Simplicity of Construction, Thoroughness in work, and Perfection as a plant protector.

It is especially adapted to the culture of all garden crops (particularly onions) and nursery stock that require careful hand cultivation in the early stages of their growth. I have placed the hoe in the hands of two of my foremen the past season, to give it a fair trial. Their report is so favorable that I think it will be decidedly for the interest of my customers to put it in my catalogue. They lay especial stress on hard ground, when it takes hold of the weeds far better than the common wheel hoe, which is too apt to slide over them. I shall order three for my own use, and I notice that several seedsmen and nurserymen have ordered from one to three for use on their own grounds after a trial of its merits.

The Frame and Wheel are made of the best cast iron. The Knives are the best cast steel. They cut the weeds only on the inside of the knife, so that any person can run it within one-half inch of the rows without injuring the smallest or tenderest plant. It will do the work of six men with the hand hoe. Boxed and sent to any address, $5 50.

NOTE.—*A*, The Cast Steel Knife, corrugated and capable of being set at any angle desired. *B*, The grooves in the casting, to adjust the Knife, from 7 to 16 inches. *C*, grooved casting, to raise or lower the handles, to suit the operator.

Excelsior Weeder.

A handy and efficient tool in the flower and vegetable garden to be used when weeds are small. By express, 30 cts.; mail, 40 cts.

Noyes' Hand Weeder.

This is an excellent little implement for hand weeding in bed sauce, such as onions, carrots and the like; it is especially valuable when the ground is hard or baked. It is much preferable to the bent hoop or knives, which are often used. To test its value on my own grounds I sent for a couple and put them in the hands of two of my boys to try; they liked them so well I sent for a dozen, and we have had them in use for several years, to our great satisfaction. The price of them sent by express is 40 cents each; per mail, 50 cents.

Gray's Garden Sprinkler.

Gray's Sprinkler is an excellent apparatus for distributing Paris Green. It holds over two buckets of water, has metal shelves within, so arranged as to prevent the Paris Green from settling at the bottom, while the motion given when walking will keep it well mixed with the water. Five acres can be gone over in one day with a saving of a pound of Paris Green to the acre. The Sprinkler can be used for common garden purposes. Full printed directions sent with each. I use Gray's Sprinkler on my farms and have found it an economical investment.

Price of the Sprinkler boxed ready for transportation...$9 00

Peerless Corn Sheller.

I first saw this neat, little implement at the New England Fair and was struck with its ingenuity, simplicity and cheapness. By means of a peculiar mechanical movement it is adapted to the shelling of all kinds of corn without scattering a single kernel; it cleans every kernel from the cob without injuring the chit, which cannot be said of any other sheller in the market. It is made wholly of iron and will not clog; will shell from 8 to 10 bushels per hour, and can be operated by a boy twelve years of age. It weighs but thirteen pounds. Price, $5.00.

ATOMIZER.

This is a valuable article for showering a fine spray of carbolic acid or solution of tobacco on the leaves of plants to destroy injurious insects. It is also used to diffuse perfumes and disinfectants in the sick room, and for inhaling various medicinal preparations. The metal parts are nickel plated. Price, by mail, postpaid, 90 cts.

Glass Cutting and Glazing Tool.

Some of my workmen who use this implement find it cuts glass as readily as a glazier's diamond, and for all the purposes of cutting and setting glass is just about equally valuable, though it costs not one-twentieth as much. One I have had in use four years is now as good as ever. Sent postpaid for 40 cents.

Okra.

Salsify.

Early Paris Cauliflower.

Hercules Club Gourd.

West India Gherkin.

Green Curled Tall
Scotch Kale.

London Green Curled Endive.

Laing's Swede Turnip.

Brussels Sprouts.

Turnip Rooted Celery.

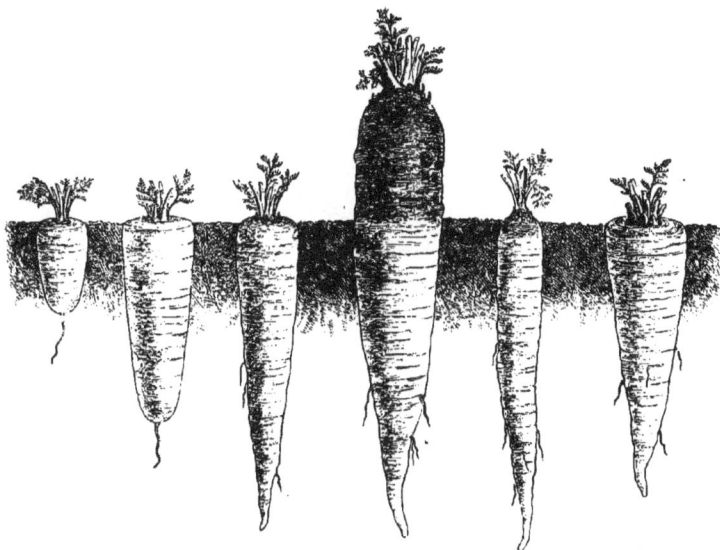

Beginning at the left hand, the smallest carrot is the Scarlet Horn, then follow Short Horn, Long Orange, White Belgian, Altringham, and Improved Long Orange, in their order.

Round Early Parsnip.

Kohl Rabi.

The above Tomatoes I have had engraved from photographs taken from specimens grown on my own grounds. No. 1 is the Mammoth Chihuahua ; No. 2, Trophy ; No. 3, Essex Early ; No. 4, Tilden ; No. 5, Canada Victor ; No. 8, Boston Market ; No. 9, General Grant ; No. 12, Orangefield.

Dwarf Curled Green Kale.

Maurandia.

Tropæolum.

Cobea Scandens.

Balloon Vine.

Abronia Umbellata.

Scypanthus.

Ipomea.

Thunbergia.

Ice Plant.

Nolana.

Ten Weeks Stock.

Viscaria.

Quilled German Aster.

Heddewig's Pink.

Gaillardia.

Pansy.

Hyacinth Larkspur.

Sanvitalia.

Primula Auricula.

Double Pot Marigold.

Nemophila.

Delphinium Formosum.

Delphinium Candelabrum (Larkspur.)

Double Portulacca.

Mimulus, Hybridus.

Spotted Rose-flowered Balsam.
(Half natural size.)

Bryonopsis Laciniata.

Double Zinnia.

Calceolaria.

Convolvulus Minor.

Pompon Bouquet Aster.

Potentilla.

Buchanan's Hybrid Petunia.

Tritoma Uvaria.

Salpiglossis.

Amaranthus Melancholicus.

Rodanthe Maculata.

Leptosiphon.

Globe Amaranth.

Lantana.

Gloxinia.

Coleus.

BULBOUS PLANTS.

Narcissus.

Single Tulip.

Crocus.

Hyacinth.

Gladiolus.

Double Buttercup.

Dahlia.

Double Tulip.

Japan Lily.

☞ For Novelties in Flowers see Pages 49 and 50. ☜

CHOICE VARIETIES OF FLOWER SEEDS.
Divided into Annuals, Biennials and Perennials.

For the convenience of my customers I have this year divided my list of Flowers into Annuals, Biennials and Perennials. I have aimed to include in my collection of Flower Seed standard varieties, with the addition of such kinds as have proved a real acquisition in floriculture. Many varieties that are usually advertised separately, I have thought best to include in a single package under the name of "mixed varieties." My three warrants I also throw around my flower seed, for it is my design that they shall be equally reliable with my vegetable seed. Let me remind my friends that flower seed, being for the most part very small in size, require extra precaution in the preparation of the soil, depth of planting, and protection from extremes of cold and wet. Do not, as a rule, plant in the open ground before the weather has become settled; better wait until the middle of May. Before planting, the soil should be made very fine and be well enriched. Then, seed the size of sweet peas may be planted three-quarters of an inch to an inch deep, and the very smallest seed should be planted barely under the surface, having fine earth sifted over them and slightly packed with the hand or a strip of board pressed upon it. It will keep the moisture in and facilitate vegetation if a newspaper is spread over the surface, after planting, and kept down with stones for two or three days. Thin out the plants when very small with a bold hand, and after they have attained to the height of two inches, thin again boldly. Give each plant plenty of room, according to its habit of growth; a very common mistake is to crowd too much. Special rates when large amounts are ordered by Clubs. Terms to Dealers on application.

ANNUALS, OR PLANTS THAT BLOOM THE FIRST YEAR AND THEN PERISH.

No.	ANNUALS.	Price per Pk'ge
1	**Abronia Umbellata** (See Plate.)	10
	Handsome, trailing plants, with clusters of sweet-scented flowers resembling the Verbena. Fine for baskets or for the garden. Sow seed under glass and set the plants eighteen inches apart. Bloom from August until October.	
	Acroclinium.	
	The most beautiful of the everlasting flowers. For Winter bouquets, cut as soon as they begin to expand. In a garden of more than two hundred varieties of flowers planted at the same time, this was the first to bloom. Half-hardy.	
2	**Acrolinium Roseum.** Bright *Rose*	10
3	——, **Alba.** Pure *White*	10
4	**Adonis Flower**	5
	A hardy plant with very pretty foliage, producing bright *red* flowers. Set plants one foot apart. Grows about one foot high and blooms for a long time.	
5	**Agrostemma, Mixed**	5
	Very pretty, free-blooming, hardy annuals, growing about one foot high, making a showy bed and fine for cutting. Can be easily transplanted, and should be set five or six inches apart.	
6	**Ageratum, Mixed**	5
	Suitable for beds and especially nice for cutting. Grows about one foot high, and continues in bloom all summer. Set six inches apart.	
	Alyssum.	
	Very fragrant, free-flowering, pretty plants for beds, edging or rock-work. Much used for bouquets on account of their delicate, honey-like fragrance.	
7	**Alyssum, Sweet.** *White*, very sweet, and blooms freely all summer. Hardy	5
	Amaranthus.	
	Ornamental foliaged plants producing a striking effect as centres of flower beds or mixed in with flowers. They are most brilliant on poor soil. Half-hardy.	
8	**Amaranthus Bicolor Ruber.** (*New.*) *Very fine*; the lower half of a leaf a *fiery red scarlet*, the upper half *maroon*. It is sometimes *tipped with yellow*; and to be unsurpassed by any ornamental leaved plant	10
9	——, **Tricolor,** (*Joseph's Coat.*) Leaves *scarlet, yellow* and *green*; beautiful; two feet.	5
10	——, **Melancholicus** (See Plate.) *Blood-red* foliage of brilliant beauty when lighted by the sun. Fine for ribboning or grouping. From Japan, one and a half feet.	5
11	**Amaranthus, Salicifolius.** This comparatively new annual is exceedingly beautiful, both in form and color. As a foliage plant it is one of the finest in cultivation. Branches of a pyramidal form, two or more feet in height. Leaves long, narrow and wavy, changing in color from a bronzy green to a brilliant scarlet, as the plant attains age. From its graceful appearance, this is sometimes called "Fountain Plant."	15

No.	ANNUALS.	Price per Pk'ge
12	**Ammobium Alatum**	5
	A very desirable *white* Everlasting Flower. Showy for the border and makes very pretty dried flowers. Hardy; two feet.	
13	**Anagallis, Mixed**	5
	Beautiful trailing hardy annuals about six inches high, and very desirable for small beds, edging, baskets, &c.	
14	**Asperula Azurea.**	5
	This is a new flower, of recent introduction. A profuse-flowering, hardy annual of Dwarf habit, with *fine foliage* and *deliciously fragrant, bright sky blue flowers*; continues in bloom till frost. Well adapted for shady places.	
	Aster.	
	Very beautiful and popular hardy annuals, growing from ten to eighteen inches high. For profusion of flowers and richness of display, the Asters are unrivalled. Sow the seed early in the spring under glass or in pots in the house, and transplant into rich soil, about one foot apart.	
15	**Aster, Coppery Scarlet.** Dwarf Chrysantho-flowered. *New.* (For description see novelties.)	15
16	——, **Dwarf Turban.** *New.* Very dark nearly black foliage. Flowers very handsome carmine mixed with white.	15
17	——, **Trophy.** *Mixed Varieties.* As hardy as the old Globe Asters; profusely branched, forming fine self-supporting globular bushes, studded with beautiful symmetrically formed flowers. The habit of the plant is extremely handsome, its height about twenty inches and perfectly constant. All the varieties are double, brilliant and true in color	10
18	——, **Dwarf Fiery Scarlet.** *New.* A new variety, highly recommended for the brilliancy of its color, not before known among Asters	25
19	——, **"Prince of Novelties."** *New.* This splendid novelty is strikingly distinct, and must prove a great acquisition. The outer ring of the flowers is of a bright crimson lake, the inner ring splashed irregularly *with rose*; centre, *pure white*	25
20	——, **Dark Purple-red,** Peony Flowered; *magnificent.*	20
21	——, **Black-Brown,** Peony Flowered; very beautiful color	20
22	——, **"Goliath."** An acquisition of extremely vigorous growth, forming large bushes from two to two and a half feet in height, *profusely* covered with flowers *immense* in size, measuring upwards of five inches in diameter, and very double. The seed which I offer is a fine mixture of *white, rose, dark, blue, ash grey and crimson*	20
23	——, **Mixed.** Flowers very globular and extremely pure in color	20
24	——, **Washington.** *Peach-bloom.* Flowers very globular and extremely delicate in tint. Seed scarce	20

No.	ANNUALS.	Price per p'k'ge
25	Aster, "The Queen's Cockade." This gem differs in habit from any other kind of Aster. The plant is of robust growth with self-supporting habit, profusely branched, representing a perfect globe of great regularity, about one foot high, densely studded with extremely double flowers of a *bright deep satiny rose*, having a *well defined, pure white center*	15
26	—, "The Meteor." A beautiful novelty. The plant is completely studded with small, globular, extremely double flowers, *brighterimson* in color. *Extremely showy and brilliant.* Fine	10
27	—, German, Double Quilled and Striped. (See Plate.) Twenty splendid varieties mixed	10
28	—, Globe Flowered, Double German. Twelve splendid varieties mixed	10
29	—, Boltze's New Dwarf Bouquet, about eight inches high, are very double, rich and free-flowering; very fine for pots or beds. Six splendid varieties mixed	15
30	—, Truffaut's French Peony-Flowered Perfection. Very large and splendid. Perfect in form, size and fullness of flower. One of the very best of the Asters. Eighteen splendid varieties mixed	10
31	—, Pyramidal, Large and Double. These are similar in style to the Peony variety, but more branching, with splendid, large, extra, double flowers. Very showy. Sixteen varieties mixed	10
32	—, Hedgehog, (or *Needle*), with white centre. Six splendid varieties mixed	10
33	—, Giant Emperor. A tall, strong growing variety, with *immense* flowers. Flowers few, but very double and often four inches in diameter. Twelve fine varieties mixed	15
34	—, Dwarf Double. A very desirable variety, of a handsome, compactly branching, bouquet-like form; the best for dwarf groups. Several varieties mixed	15
35	—, Dwarf Pompon Bouquet. (See Plate.) A beautiful Aster with neat, pompon flowers. Many very fine varieties mixed	15
36	—, Above varieties mixed	15
37	Azalea. (*Finest varieties mixed.*) One of the most ornamental of greenhouse shrubs and *admirably adapted for a window plant.* It flowers with great profusion. The roots being very fine, it requires a light soil. The seed, which I send out, I have imported from Germany, from one of the most celebrated of growers.	25
38	Balloon Vine. Ornamental climbing plant, remarkable for an inflated membranous capsule, (*white* in color) from which it derives the name of Balloon Vine. Suitable for the greenhouse or out-door decoration. Half-hardy.	10
	Balsams. Most magnificent, and popular half-hardy annuals, from one and one-half to two feet in height. Sow the seed early in frames, and transplant to a well enriched border, one foot apart.	
39	Balsams, Double Yellow. (New.)	15
40	—, Camelia-flowered, Carmine. (New.) Splendid new luminous color	20
41	—, Double Mixed Camelia Flowered	10
42	—, Spotted, Rose-Flowered, Mixed. (See Plate.) Very large flowered, double and with perfectly formed flowers; from six to eight fine varieties, all spotted with white	15
43	—, Dwarf Camelia Mixed. From eight splendid double varieties; one-half foot	10
44	Bartonia Aurea. A showy, hardy annual, growing about two feet high, bearing very *brilliant yellow* flowers. Thin plants to six inches apart.	5
45	Dell's Ornamental Dwarf Beet. Leaves peculiarly rich in color; highly ornamental as a foliaged plant.	5
46	Bocconia frutescens. Beautiful foliaged greenhouse plant, suitable for lawns in Summer.	15

No.	ANNUALS.	Price per P'k'ge
	Browallis. Very handsome, profusely blooming plants, covered with rich and strikingly beautiful flowers; grows freely in rich soil. Very desirable grown in beds; one and one half feet. Half-hardy.	
47	Browallia, Roezli, *New.* (For description see novelties.)	20
48	—, Blue, with white centre. Very brilliant and beautiful, one and one-half feet	10
49	—, Mixed. One-half foot	10
50	Bryonopsis Laciniata. (See Plate.) An *elegant* climber, with *Ivy-like foliage* and *brilliant scarlet fruit* exquisitely marked with white. Ten feet. Half-hardy.	10
51	Canary Bird Flower. A highly ornamental creeper, with exceedingly beautiful *yellow* fringed flowers and finely divided foliage; a general favorite. Half-hardy.	10
	Candytuft. A well known hardy annual, very useful for bedding and bouquets, and also for pot culture.	
52	Candytuft, New Carmine	25
53	—, White	5
54	—, Purple	5
55	—, Crimson	5
56	—, Fragrant	5
	Catch-fly. A hardy annual about one foot high:—produces brilliant flowers in great profusion in early summer.	
57	Catch-fly, Red	5
58	—, White	5
	Centranthus. Exceedingly pretty, free-flowering plants, from six inches to one foot high, and very effective in beds or borders.	
59	Centranthus, Rose	5
60	—, White	5
	Chrysanthemums. Handsome hardy annuals about two feet high, blooming from July to October.	
61	Chrysanthemums, Double White	5
62	—, Double Yellow	5
63	—, Mixed	5
	Cineraria. A well known greenhouse favorite. Remarkable for its free-flowering habit and beauty of foliage. May be had in splendid bloom throughout the greater portion of the year.	
64	Cineraria, Fine Mixed	25
	Clarkia. A very pretty, hardy annual, about eighteen inches high blooming from June to September. Set six inches apart.	
65	Clarkia, Rosea	5
66	—, White	5
67	—, Purple	5
68	—, Mixed	5
69	Clianthus Dampieri. (*Glory Pea.*) One of the most attractive plants in cultivation. A *shrubby climber*, with *neat foliage* and *drooping clusters* of large, rich, scarlet, pea-shaped flowers, 3 inches in length. Each flower is beautifully marked with a black, cloud-like blotch. It requires a dry, warm soil and should be watered but sparingly. Remove to the house when cold weather comes.	25
	Clitoria. Splendid, free-flowering, greenhouse climbers, with large, elegant, pea-shaped flowers. Particularly adapted for training on trellis-work, wire globes, &c.	
70	Clitoria, Cœlestis. *Sky-blue*; from East Indies	10

No.	ANNUALS.	Price per P'k'ge
	Cobea Scandens. (See Plate.)	
	A magnificent, climbing plant of rapid growth, producing large, purple, bell-shaped flowers; adapted for out door growth in Summer, also for the house and greenhouse. The seeds are apt to rot in the open ground, hence should be started in the house. Place the seed edgewise in planting.	
72	Cobea Scandens..............................	10
73	—, Alba. White flowered variety of that famous climber.	25
	Cockscomb.	
	Very singular and attractive hardy annuals, about two feet high.	
74	Cockscomb, Japonica alba, New, (For description see novelties.)............................	15
75	—, Cristata Variegata. New and beautiful. Gold and Crimson Variegated.................	20
76	—, New Japan..............................	15
77	—, Crimson. Tall, very fine (pure.)...........	15
78	—, Dwarf Mixed. Very beautiful...............	10
79	—, Feathered. New and very fine for bouquets......	10
80	**Coleus.** (New and finest varieties Mixed) (See Plate.)	35
	These gorgeous colored plants with their variegated foliage, are deservedly called the most popular plants in cultivation. Their leaf markings are rich and varied. No garden, basket, vase nor collection of flowers is complete without them. Start the seed in a hot bed or pot in the house and transplant when warm weather comes.	
81	**Collinsia, Mixed**	5
	Beautiful, hardy annuals, very pretty when grown in a mass. Thin plants to three inches apart.	
82	**Collomia, Scarlet.**	5
	A very pretty annual, producing heads of bright red flowers.	
	Convolvulus.	
	Beautiful and showy half hardy climbers, producing an abundance of rich colored flowers. Set plants one foot apart. Blooms from July until autumn.	
83	**Convolvulus, Minor, Dwarf Mixed.** (See Plate.)..	5
84	—, Striped. Blue, beautifully striped with white...	5
85	**Convolvulus, Major.** (Morning Glory.) Fine varieties mixed. A well-known, beautiful, climbing annual, suitable for covering arbors, trellises, &c. Blooms from July until autumn..................	5
	Coreopsis. (Calliopsis.)	
	One of the most showy of all annuals, the colors being so striking as seldom to be passed without remark.	
86	**Coreopsis, Drummondii.** Large yellow flowers, with crimson centre............................	5
87	—, Mixed varieties. Yellow and brown and velvety-brown............................	5
88	**Cosmidium, Burridge's**....................	5
	An elegant annual, growing about two feet high, with rich, brilliantly colored flowers.	
	Cypress Vine.	
	Most beautiful and popular tender climber, with very delicate fern-like foliage and numerous flowers.	
89	Cypress Vine, Scarlet. Very brilliant...........	5
90	—, White..................................	5
	Delphinium.	
	Flowers remarkable for their great beauty, diversity of shades and striking appearance. Hardy.	
91	**Delphinium Candelabrum.** (See Plate.) A new and elegant variety of Larkspur. The branches are beautifully curved, (pointing upwards,) diminishing in length as they approach the top of the center spike, thus giving the plant the appearance of a candelabra. Flowers brilliant and varied................	15
92	**Erysimum Peroffskianum.** (See plate.).......	5
	A very showy, hardy annual about eighteen inches high,—bearing spikes of deep orange-colored blossoms. Blooms from June to September.	
	Eschscholtzia.	
	A very showy plant about one foot high, blooming from June until September. Produces a brilliant effect at a distance when grown in a mass. Hardy.	
93	Eschscholtzia, Mandarin. New. (For description see novelties.)................................	25

No.	ANNUAL	Price per P'k'ge
94	Eschscholtzia, Yellow........................	5
95	—, White................................	5
96	—, Tenuifolia..............................	5
97	—, Mixed................................	5
	Eternal Flower. (Helichrysum.)	
	The Eternal Flowers are very ornamental in the garden and very desirable for winter bouquets, as they will retain their form and color for years if gathered and dried when first open.	
98	Eternal Flower, Yellow......................	5
99	—, Mixed..............................	5
100	**Euphorbia, or Snow on the Mountain**........	5
	A very pretty variegated, foliage plant, leaves edged with pure white. Tender.	
101	**Flax, Crimson.** (Linum Grandiflorum.).........	5
	A beautiful, half-hardy annual, one foot high and very effective and showy for bedding purposes. Set plants one foot apart.	
102	**Fuchsia, or "Ladies' Ear Drop."** (Finest and newest varieties mixed.).............	35
	Elegant flowering plants of easy culture in pots for parlor decoration or the garden. In the garden they require a slightly shaded situation. Soil should be rich.	
103	**Gaillardia, Mixed.** (See Plate.)............	5
	Hardy annuals, universally admired for their fine display. Grow about eighteen inches high, and bloom all summer.	
	Gaura.	
	An exceedingly handsome and free-flowering plant, continuing in bloom the whole summer. Of light and graceful habit, bearing spikes of white and red tinted flowers; a profuse bloomer. Succeeds best in sandy loam. Half-hardy.	
104	Gaura, Lindheimeri. White with pink calyx; from Texas. Two feet............................	5
105	**Geranium, Fancy, Splendid Mixed**.........	25
	Gilia.	
	Early and free-flowering, hardy annuals, growing from six inches to one foot high, and very desirable for planting in masses or detached patches.	
107	Gilia, White..............................	5
108	—, Rose..............................	5
109	—, Tricolor............................	5
	Globe Amaranths. (See Plate.)	
	Tender annuals about two feet high, very ornamental in the garden. The flowers will retain their beauty for a long time if gathered and dried as soon as they are open. Start early in hot-bed, and transplant one foot apart in the border when the weather becomes warm.	
110	Globe Amaranth, White......................	5
111	—, Purple..............................	5
112	—, Variegated.........................	5
113	—, Mixed..............................	5
114	**Gloxinia,** (grandiflora).....................	50
	A superb class of greenhouse and in-door plants, producing, in great profusion, elegant flowers of the richest and most beautiful colors. This variety I send out is very choice and is one of the new varieties, with vigorous foliage and very large flowers in the best and liveliest colors.	
	Godetia.	
	Very attractive, hardy annuals of easy culture, about one foot high, flowering in July and August. Flowers of satiny texture.	
115	**Godetia, Lady Albemarle.** (New.) Plants about one foot high, branching from the bottom, and growing in a pyramidal form; flowers large, frequently measuring 3 1-2 to 4 inches across, and of the most intense glowing carmine color. The edges of the petals slightly suffused with delicate lilac. The flower are produced in such wonderful profusion and are of such brilliant color that the plants have the most brilliant appearance. It is perfectly hardy, and if sown out of doors in Autumn will bloom early in the following summer..............................	15
116	—, Mixed..............................	5
	Grasses, Ornamental.	
	The Ornamental Grasses are most desirable for bouquets both for Summer and Winter. For Winter use, cut about the time of flowering, tie up in small bunches and dry in the shade.	
117	Grasses, Ornamental, Eragrostis Brown, New, (For description see novelties.).................	10

No.	ANNUALS.	Price per Pk'ge
118	Grasses, Ornamental, Agrostis Nebulosa. The most delicate, fine and feathery of the Ornamental Grasses. Hardy	10
119	——, ——, Eragrostis Elegans, (" Love Grass.") An exceedingly pretty and highly ornamental grass. Grows one to two feet. Hardy.	5
120	——, ——, "Job's Tears." This well known variety of tropical grass is so called from the appearance of its shiny, pearly fruit, which resembles a falling tear. Half-hardy.	5
121	——, ——, Quaking Grass. This graceful shaking grass is very elegant in bouquets and may be dried and kept a long time; perfectly hardy.	5
	Gypsophila.	
	Elegant free flowering little plants, succeeding in any soil. Well adapted to rockwork and edging. Ladies will find this desirable for ornamenting their hair.	
122	——, Muralis. Beautiful, dwarf plant, neat and remarkably pretty, with starry pink and white flowers which completely cover the plant. Very fine for hanging baskets. Hardy annual, one half foot.	10
123	Hawkweed, Mixed. (Crepis.)	5
	A class of attractive hardy annuals, one foot high, of easy culture.	
124	Heliptorum Sanfordi.	10
	A new variety of everlasting flowers of great beauty; of dwarf, tufted habit, producing large, globular clusters of bright golden yellow flowers, excellent for winter bouquets.	
125	Hibiscus Africanus.	5
	A showy and beautiful, hardy annual, eighteen inches high, blooming from June to September. Set eighteen inches apart.	
126	Hollyhock. Dwarf Chinese. Showy, hardy annual variety, two and a half feet high. Start early in hot-bed and transplant one foot apart.	5
127	Hyacinth Bean. (Dolichos.)	5
	Tender, climbing annual from the East Indies, producing clusters of brilliant flowers.	
128	Ice Plant. (See Plate.)	5
	A singular-looking, tender annual with thick, fleshy leaves, that have the appearance of being covered with crystals of ice.	
	Ipomea. (See Plate.)	
	Very beautiful and popular climbers; deservedly so from the fine foliage and the brilliant and varied hues of its many flowers. Fine for covering old walls, stumps of trees, &c.	
129	Ipomea Atroviolacea, violet, bordered with pure white; superb	10
130	——, Bona Nox, Evening Glory. This very interesting plant is as its name indicates, allied to the " Morning Glory," but differs from it in choosing the evening for its time of blooming. It is also odorously fragrant. The flower is pure white and very large. Soak the seed in warm water several hours before planting.	15
131	——, Elegantissima. One of the richest of the Ipomeas; blue with intense purple centre in the form of a star, with broad, pure white margin.	10
132	——, Limbata. Blue, elegantly marked with white	10
133	——, Nil Grandiflora. A very beautiful variety from Germany	10
134	——, Coccinea. (Star Ipomea.) A beautiful, climbing, tender annual, closely allied to the Morning Glory, producing a profusion of bright scarlet flowers	10
	Jacobea. (Senecio.)	
	A very gay-colored, showy class of hardy plants, very effective for bedding. Grow about one foot high.	
135	Jacobea, Double, White.	10
136	——, Double, Dark Blue.	10
137	——, Double, Mixed	10
138	Kale, Ornamental. Four elegant varieties. Very desirable as a foliage plant	5

No.	ANNUALS.	Price per Pk'ge
	Larkspur.	
	Very beautiful, hardy annuals, producing dense spikes of flowers, which are very decorative either in the garden or when cut for vases. Set ten inches apart.	
139	Larkspur, Dwarf Ranuncull-Flowered, New (For description see novelties.)	15
140	——, Tall Double Rocket	5
141	——, Stock-flowered. (See plate.) Eight varieties mixed	5
142	——, Hyacinth-flowered. (See Plate.) A curious and very beautiful variety, strongly resembling a Double Hyacinth. Twelve fine varieties mixed.	10
143	——, Tricolor Elegans. A very double variety of very handsome colors and most beautifully striped. Two and one-half feet.	10
	These two last named varieties may be considered as great acquisitions to the garden.	
	Leptosiphon. (See Plate.)	
	The most desirable of plants for edgings; very beautiful with their numerous and many colored flowers; also suitable for rock-work, and nice for pot plants; succeed in any light, rich soil; from California. Hardy.	
144	Leptosiphon Mixed. Colors dark maroon, orange, lilac, purple, crimson, violet, golden yellow and white. Exceedingly pretty.	10
	Lobelia.	
	Strikingly pretty, profuse-blooming plants; their delicate, drooping habit and the profusion of their charming little flowers render them exceedingly ornamental. Very fine for hanging baskets. Hardy.	
145	Lobelia, Pumila Magnifica. (New.)	25
	This is by far the finest form of Lobelia in cultivation. The habit of the plant resembles the fine foliaged Pumila variety, while the flowers are of immense size, and are of the richest ultramarine blue color.	
146	——, Rosea Oculata. Rose, with white eye.	15
147	——, Erinoides. Blue.	5
148	——, Erinus, Mixed. Blue, white, and blue and white marbled.	5
	Lophospermum.	
	An exceedingly elegant and highly ornamental climber with large and handsome foxglove-like flowers; very effective for conservatory and garden decoration, and also desirable for hanging baskets; blooms the first season from seed. Half-hardy.	
149	Lophospermum Hendersonii. Flowers of rosy carmine, fine. Ten feet	15
150	Love-lies-bleeding. (Amaranthus Caudatus.)	5
	A hardy annual, three to four feet high, with pendant spikes of blood-red flowers, which at a little distance look like streams of blood. Desirable for grouping on lawns.	
151	Love-in-a-mist. (Nigella.)	5
	A curious plant about one foot high, with finely cut leaves and singular flowers. Hardy.	
	Malope.	
	Handsome, half-hardy annuals, about two feet high. Set eighteen inches apart. Well adapted to mixed borders.	
152	Malope, Mixed.	5
153	Marvel of Peru, Mixed. (Mirabilis.)	5
	The old and well-known Four o'clock. A fine plant with flowers of various colors, making a fine summer hedge when set one foot apart. Grows two feet high. The roots may be preserved like Dahlias during the winter. Half-hardy	
	Marigold. (Tagetes.)	
	Extremely showy, half-hardy annuals, one to two feet high, well adapted to garden culture, blooming profusely through the season. Set one foot apart.	
154	Marigold, African.	5
155	——, Pot. (See Plate.) This variety of marigold is well worth much praise. The flowers are large, very brilliant and double; in color varying from a deep orange to a pale lemon and have a dark maroon centre. They begin blooming very early and continue till after the heavy frosts. They sow their own seed and thus perpetuate themselves.	5
156	——, ——, French.	5
157	——, ——, Gold-striped, new and fine.	10

No.	ANNUALS.	Price per Pk'ge	No.	ANNUALS.	Price per Pk'ge

Maurandia. (See Plate.)

An elegant, half-hardy, climber, well adapted to the conservatory or trellis work in the garden. Start early in pots, transplant when the weather becomes warm. Flowers the first year from the seed and continues to bloom through the season. Desirable for hanging baskets.

| 158 | Maurandia, Barclayana. Rich, *violet* flowers. | 5 |
| 159 | —, Mixed. *Violet, white, rose* and *pink* | 10 |

Mignonette.

A hardy annual, eight inches high. A general favorite on account of its delightful fragrance. Blooms throughout the season. Sow from middle of April to middle of June. Thin to six inches apart. Mignonette is most fragrant on poor soil.

160 **Mignonette, Miles' Hybrid Spiral.** This variety is far superior to any other in cultivation, the habit being dwarf and branching, with spikes often attaining a length of from 8 to 14 inches. By pinching the side shoots the centre spike attains a length of from 18 to 21 inches. The odor of this variety is superior to any other in cultivation. It is much hardier, and well adapted for market purposes.

161	—, Sweet. ....per ounce 25 cents.	5
162	—, Grandiflora. An improvement on the old variety in size	5
163	—, Victoria. *New.* Flowers unusually brilliant.	20

Momordica.

Trailing plants with ornamental foliage and *golden yellow* fruit which, when ripe, opens, disclosing its seeds and brilliant *carmine* interior. Planted on rock-work or stumps of trees and allowed to ramble, it produces a very striking effect. Half-hardy.

| 164 | Momordica Charantia, or Balsam Pear | 5 |
| 165 | Morning Glory. (See Convolvulus Major.) | 5 |

Morning Bride. (*Scabiosa.*)

A class of very pretty annuals, from one to two feet high—suitable for bedding or bouquets. Hardy.

166	Mourning Bride, Mixed	5
167	—, Double. *Cherry color.* (*New.*) An acquisition	10
168	—, Dwarf. Six splendid varieties, mixed.	5

Nasturtium. Hardy Annual.

169	Nasturtium, Tall mixed	5
170	—Purplish-violet. (*Tom Thumb.*) *New.*	15
171	—"Ruby King" *Pure pink shaded with crimson.*	15
172	—"Spotted King" *Bright yellow blotched with chocolate.*	15
173	—, Scheuerianum. *Straw colored striped with brown. Very beautiful.*	10
174	—, Spitfire. *New.* Very fine; bright *fire-red.* Flowers very freely and makes a strikingly showy appearance.	15
175	—, Dunnett's New. *Orange.*	5
176	—, Atropurpurea. *Dark blood crimson.*	5
177	—, Coccineum. Brilliant scarlet.	5
178	—, Dwarf Mixed.	5
179	—, Dwarf Scarlet.	10
180	—, Rose.	10
181	—, White. (The pearl.)	10
182	—, Yellow.	10
183	—, King of Tom Thumb's. *Deep scarlet* blossom, *bluish green foliage,* new and fine.	10

Nemophila. (See Plate.)

Charming, hardy, low annuals, producing an abundance of extremely delicate and beautiful flowers. Very useful for bedding or for pot culture. Sow early in pots and transplant into a cool, rather moist situation.

| 184 | Nemophila Mixed | 5 |

185 **Nolana. Mixed.** (See Plate.)

Very pretty, trailing hardy annuals, fine for rock-work, hanging baskets, or for bedding. Select light rich soil.

186 **Oxlip, Sweet Scented, Mixed.** Of beautiful colors ... 15

Pansy. See Heartsease, in list of Perennials.

Parsley.

I would recommend Curled Parsley as fine for edgings for the flower garden, fine for vase bouquets, and particularly desirable for flowers arranged in flat dishes.

| 187 | Parsley, Dwarf Curled. | 5 |
| 188 | —, Fern Leaved. A most beautiful thing. Invaluable as a decorative plant. Resembles a beautiful moss. | 10 |

| 189 | Perilla Nankinensis. | 5 |

A half-hardy annual, with beautiful *dark purple foliage* forming a delightful contrast with the lively green of the other plants in the garden or conservatory.

190 **Phaseolus.** (*Scarlet Runner Bean.*) | 5

A popular climber, with spikes of showy *scarlet, white* or *variegated* flowers. Extensively grown to cover arbors and to form screens; of very vigorous and rapid growth.

Phlox.

A most brilliant and beautiful hardy annual, about one foot high, well adapted for bedding, making a dazzling show through the whole season. It succeeds well on almost any soil.

191 **Phlox, Drummondi, Alba Oculata Superba,** New. (For description see novelties.) | 20

192	—, Victoria, New. (For description see novelties.)	20
193	—, Heynholdi Alba. *New.* The flowers of this beautiful new variety are wholly snow-white. It is, indeed, the purest white Phlox yet raised, well adapted for pot culture.	30
194	—, Drummondi Grandiflora Splendens. (*New.*) Flowers large, handsomely rounded and of great substance; *color vivid with a pure white disc;* habit of growth free and robust.	10
195	"Fireball." (*New.*) A splendid new dwarf variety. Grows in large robust bushes quite covered with *brilliant-red* flowers till late in autumn, giving the bush the appearance of a Fireball.	25

196	Phlox, Pure White.	10
197	—, —. Bright Scarlet.	10
198	—, —. Crimson, *striped with white, very beautiful,*	15
199	—, —. Splendid, *red with white eye.*	10
200	—, —. All Colors Mixed.	5

Pinks.

Most beautiful and highly prized, hardy plants, growing from one to two feet high. No garden is complete without them, as they keep up a brilliant display, almost the whole season. Start early in pots, and transplant six to ten inches apart.

201	—, Carnation. (See list of Perennials.)	10
202	—, Carnation, Dwarf Fiery Red. (See list of Per.)	35
203	—, Picotee. (See List of Perennials.)	25
204	—, Heddewig's. (See list of Biennials.)	10
205	—, Chinese. (See List of Biennials.)	5
206	—, Laciniatus, Finest Double Mixed. Magnificent, double flowers; very large and beautifully fringed. Saved only from the finest double flowers and most beautiful colors.	15

207 **Poppy, Double Mixed.** (*Papaver.*) | 5

Brilliant and showy, hardy annuals, about two feet high, fine for back ground or shrubbery.

Portulacca.

Very popular; low growing, plants; making a most brilliant display in the garden, and very suitable for borders or edging. Sow early, in warm, light soil and thin plants to four inches. Hardy.

208	Portulacca, all colors Mixed.	5
209	—, Scarlet.	5
210	—, Crimson.	5
211	—, White.	5
212	—, Yellow.	5
213	—, Large flowered Double. (Bernary's Best.) (See Plate.)	20

214 **Ricinus Major. Castor Oil Plant.** | 5

A highly ornamental, half-hardy annual, growing from four to six feet high, presenting quite a tropical appearance. Select warm, dry soil, and plant six feet apart.

Rodanthe. (See Plate.)

A most beautiful and charming everlasting flower. The flowers, when gathered as soon as they are opened, are very desirable for winter bouquets, retaining their brilliancy for months. Half-hardy annual.

| 215 | Rodanthe, Maculata. | 10 |
| 216 | Salpiglossis, Mixed. (See Plate.) | 10 |

Very beautiful, rich, half-hardy annuals, of varied colors, one to two feet high. Start early in the hot-bed and transplant to light, warm, rich soil. Blooms from July to September.

Salvia.

Very ornamental, plants two feet high, producing tall spikes of gay flowers. Sow early in hot bed and transplant two feet apart. Half-hardy.

No.	ANNUALS.	Price per Pk'ge

217 Salvia, Grandiflora Bicolor. (*New.*) The foliage is varigated with white, and the flowers are *white and rose with scarlet tip.* ... 25

218 —, **Mixed.** ... 5

219 —, **Coccinea.** Splendid scarlet ... 10

Sanvitalia. (See Plate.)

Beautiful, dwarf-growing, free-flowering plants, very suitable for small beds or rock-work. Hardy annuals.

220 Sanvitalia, New, Double. Covered with dense masses of perfectly double flowers. This variety is considered by an experienced grower of rare flowers, as "without doubt the only dwarf, compact plant, of a *yellow color,* suited to beds and masses of low growth" ... 10

221 Schizanthus, Mixed. ... 5

Pretty, tender annuals, one to one and one-half feet high, blooming from August to October. Very pretty for pot culture.

222 Scrophularie Chrysantha, New. For des. see novelties. ... 15

Scypanthus.

A very ornamental, free-flowering climber, with curious *yellow* flowers, producing a fine effect trained against verandahs, trellises, &c. Half-hardy.

223 Scypanthus, Elegans. *Yellow,* from Chili. ... 10

224 Sensitive Plant. (*Mimosa Sensitiva*) ... 5

A pretty, curious annual, being so sensitive that the leaves close together by the slightest touch. Tender annual.

Stocks.

Half-hardy annuals, producing splendid spikes of very rich and beautiful flowers of delightful fragrance. For early flowering sow early in spring in pots or in the hot bed, and transplant one foot apart. Bloom from June until November.

225 Stocks, New Perpetual Flowering, Double White. New. For description see novelties. ... 25

226 —, **Dwarf German.** (*Finest Mixed.*). ... 10

227 —, **Ten Weeks, Double Mixed. (See Plate.)** ... 10

228 Stocks, Dwarf, Large Flowering. One of the finest stocks in cultivation. Very double and of a rich *dark crimson* color. ... 20

Sunflower. (*Helianthus.*)

The most beautiful and ornamental of this well-known class of plants, growing about four feet high, and producing very large double flowers. Hardy.

229 Sunflower, Dwarf Double ... 5

230 —, **Variegated.** Flowers and end of stock variegated. ... 10

231 —, **Globosus.** The finest of all sunflowers for ornament. The plant is middling sized, flowers very large, completely double, of a bright golden yellow. ... 5

232 Swan River Daisy. (*Brachycome*) ... 10

Very pretty, free-flowering, dwarf-growing annuals, well adapted to edgings, rustic baskets, or for pot culture.

Sweet Sultan.

Showy, hardy annuals, one to two feet high, succeeding well in any soil.

233 Sweet Sultan, Mixed. (*Centaurea.*). ... 5

234 —, **Yellow.** Much may be said in praise of this the most beautiful of the Sweet Sultan family. The flowers are remarkably long-lived, which, with its beautiful fragrance and golden color, renders it very desirable for bouquets. ... 5

235 Sweet Clover. Valuable for its fragrance. ... 5

Sweet Peas. (*Lathyrus Odoratus.*)

Very ornamental, hardy annuals, desirable for their delightful fragrance and beauty. Fine for covering fences or walls, or for growing in little clumps supported by sticks. By picking off the pods as soon as they appear, the blossoms may be continued the whole season.

236 Sweet Peas, White. ... 5

237 —, **Black.** ... 5

No.	ANNUALS.	Price per Pk'ge

238 Sweet Peas, Scarlet. ... 5

239 —, **Scarlet Striped with White.** ... 5

240 —, **All Colors Mixed.** Per ounce 25 cents. ... 5

241 Tassel Flower, Scarlet. (*Cacalia.*). ... 5

A beautiful, half-hardy annual, with small, tassel-like flowers, blooming profusely from July to October.

Thunbergia. (See Plate.)

Very ornamental, trailing or climbing, half-hardy annuals, admirably adapted for trellises or rustic work or for the conservatory. A great acquisition for hanging baskets. Start early under glass. Tender annual.

242 Thunbergia Coccinea. New. A deep scarlet variety of this beautiful free-flowering climber. ... 25

243 —, **Mixed.** Flowers white and salmon, with rich maroon centres. ... 10

244 Tropaeolum Mixed. ... 10

Half-hardy annuals, very ornamental, and easily cultivated as climbers, producing an abundance of richly colored flowers. These are selected from the finest English varieties.

245 Umbilicus Sempervivum. (*New.*) From Kurdistan, a small unique form of superviyum; the second year it throws up a large nubel of beautiful blood-red flowers; the whole plant changes then from green to red. A capital plant for carpet gardening. The plant is a beauty in its way, and it would be impossible to say too much of it. Its hardiness has not yet been tested, but probably it is hardy throughout the Middle States. Sow in boxes or pans, and plant out in the following summer, its flowers attaining a height of six inches ... 15

246 Venus Looking-Glass, Mixed. ... 5

A very pretty, hardy, annual succeeding well in any soil. Grows about one foot high, and is well adapted to borders or edgings.

Verbenas.

Well-known and universally popular bedding plants, blooming all summer. May be treated as half-hardy annuals. Sow the seed early under glass and transplant one foot apart. Tender perennials.

247 Verbena, Fine Mixed ... 10

248 —, **Finest Mixed.** ... 15

Virginian Pigmy Stock.

Extremely pretty, profuse-flowering, little plants, remarkably effective in small beds, baskets or edgings. Hardy annuals.

249 Virginian Stock, White. One-half foot. ... 5

250 —, **New Rose.** One-quarter foot. ... 5

251 Viscaria, Mixed, or "Rose of Heaven." (SEE PLATE.) ... 5

Very pretty, profuse-flowering, half-hardy annuals, producing a fine effect in beds or mixed borders, and growing readily in any soil.

Xeranthemum.

Very showy, free-flowering everlasting flowers, valuable for winter bouquets. Hardy annuals.

252 Xeranthemum Annum Superbissimum. New. For description see novelties. ... 10

253 —, **Double, White** ... 5

254 —, **Purple** ... 5

Zinnia. (See Plate.)

A most splendid class of hardy annuals, succeeding well in any soil and making a very brilliant show. Start early in pots or under glass and transplant one foot apart. The same flowers will retain their beauty for weeks and a profusion will be produced until frost.

255 Zinnia, Tall. Finest varieties double mixed. ... 10

256 —, **Double White** ... 15

257 —, **Haageana.** Comparatively New; of dwarf, branching habit; each petal yellow flushed with orange. An exceedingly valuable plant for flower beds, edgings or borders. ... 10

258 —, **Double Sulphurea Striata** New. Sulphur colored, striped with scarlet. Very showy and beautiful when distinct in its colors. ... 20

BIENNIALS, OR PLANTS WHICH LIVE AND GENERALLY BLOOM TWO YEARS.

No.	BIENNIALS.	Price per P'k'ge	No.	BIENNIALS.	Price per P'k'ge
	Alonsos.		270	**Foxglove, Mixed.** (*Digitalis.*).........	5
	A very ornamental bedding plant ; flowers freely from June till the frost—a half hardy biennial.			A hardy biennial, growing three to four feet high and very ornamental in the garden or amongst shrubbery, as it produces tall spikes of blue and white, bell-shaped flowers.	
259	**Alonson Grandiflora,** (large flowered,) deep *scarlet*, two feet high........	5			
260	——, **Warszewiczi,** bright *crimson*, from Chili, one and a half feet high........	5		**Heartsease, or Pansy.** (See Plate.)	
	Canterbury Bells.			A well-known and universal favorite ; properly a biennial, but may be propagated by cuttings or by dividing the roots. It blooms early the first season and produces a profusion of brilliant flowers from early spring until winter. It will thrive well anywhere, but prefers a moist, shady situation.	
	Well known biennials, growing about one foot high, producing beautiful bell-shaped flowers. Set six inches apart.				
261	**Canterbury Bells, Blue Single.**........	5			
262	——, **White Single.**........	5			
263	——, **Double Mixed.**........	10	271	**Heartsease, or Pansy, International Paris.** New. (For description see novelties)........	30
264	**Honesty.** (*Lunaria.*)........	5	272	——, **Fine mixed.**........	5
	Blooms in May and June. The flowers are succeeded by singular, semi-transparent seed-vessels that are quite ornamental and may be kept for a long time.		273	——, (*Pure Yellow, Large Flowered.*) The brilliancy and beauty of this Pansy make it a great favorite.	20
265	**Humea Elegans**........	15	274	——, (*Extra choice mixed.*) These varieties are very superior.	15
	A magnificent, showy, half-hardy biennial, four to eight feet high, blooming the second year through the summer and autumn. Very ornamental in the garden.		275	——, (*Finest, very large stained.*)........	25
	Ipomopsis.		276	——, **King of the Blacks.** Deep *Coal Black.*..	20
	Most beautiful plants with spikes of dazzling flowers.		277	——, **Pure White.**........	20
266	**Ipomopsis,** *orange*, from California ; three feet......	10	278	——, **Odier, or Five Blotched.** *A new and beautiful Price Pansy,* of great variety of color and markings, each petal being most beautifully blotched or marked. The seed I send out is from the celebrated Beuary, and is extra choice and true........	25
267	——, **Elegans,** *scarlet*........	10			
	Pinks.				
268	**Pink, Heddewig's Double Mixed.** (See Plate.) Large flowers, three inches in diameter, of beautiful and rich colors, often finely marked and marbled......	10	279	——, **Emperor William.** One of the most valuable of the large growing pansies. Flowers of a rich ultramarine, with a well-defined eye. The large blooms are borne well above the foliage........	25
269	——, **Chinese.**........	5			

PERENNIALS, OR PLANTS WHICH LIVE MORE THAN TWO YEARS.

No.	PERENNIALS.	Price per P'k'ge	No.	PERENNIALS.	Price per P'k'ge
280	**Aconitum.** (*Monkshood.*)........	5		**Cineraria.**	
	A hardy perennial, grows well in any good soil, even when in the shade.			A well-known greenhouse perennial. Remarkable for its free-flowering habit and beauty of foliage. May be had in splendid bloom throughout the greater portion of the year.	
281	**Alyssum, Saxatile.** *Yellow*, extremely showy. A hardy perennial........	5	290	**Cineraria, Maritima.** Flowers *yellow*, leaves large and silvery; an ornamental foliaged plant, fit for edgings, in which case it should be kept from flowering. It forms a fine contrast, in ribboning, with Perilla Nankinensis. One and one-half feet........	10
282	**Aristolochia, Mixed.**........	15			
	Highly ornamental and attractive climbers, with curiously shaped flowers of the most varied and beautiful colors. The flowers resemble a Dutchman's pipe. Hardy perennial.			**Clematis.**	
283	**Bachelor's Button** (*Centaurea.*)........	5		Beautiful, hardy climbers, unrivalled for covering arbors, fences, verandahs, &c; will succeed in any good garden soil.	
	A showy, hardy annual, about two feet high—succeeding well in any soil. Set six inches apart.		291	——, **Cirrhosa.** Perfectly hardy, a very rapid climber, literally covering itself with large bunches of *white, sweet-scented* flowers. Twenty-five feet.........	25
284	**Baptisia Australis**........	5	292	——, **Graveolens.** New. (For description see novelties.)	20
	A handsome plant of the easiest culture ; flowers *blue* and *white.* Hardy perennial two feet.		293	——, **Pilcheri.** New. (For description see novelties.)	20
285	**Bellis Perennis.** (*Double Daisy.*)........	10		**Columbine.** (See plate.)	
	A favorite perennial for the border or for pot culture. Set plants six inches apart.			A well known, showy, hardy perennial, about two feet high, blooming in May and June.	
	Calceolaria.		294	**Columbine, Aqilugia truncata,** *New.* (For description see novelties.)........	15
	Flowers highly decorative ; very desirable, indeed invaluable, for the house, greenhouse and the garden. Seeds should be started in pots, but not under glass. Half-hardy perennials.		295	——, **Mixed**........	5
			296	——, **California.** A California species, large and handsome, the color being of a waxy yellow. Remarkably fine........	25
286	**Calceolaria, Tigridus.** (See Plate.) A new spotted variety, extra fine........	25	297	**Dahlia, Mixed.**........	15
	Callirhoe.			Seeds saved from very fine named sorts mixed, from France. Half-hardy perennial.	
	Beautiful, free-flowering plants, beginning to bloom when small and continuing throughout the summer and fall ; excellent for beds or masses when sown thick. Hardy annual.			**Datura.** (*Trumpet Flower.*) —	
287	**Callirhoe, Pedata.** Rich *purple crimson*, with white eye, two feet........	10		A showy, half-hardy perennial, producing large, sweet-scented, trumpet-shaped flowers. The roots should be removed to the cellar in autumn. Two and a half feet.	
288	——, **Involucrata.** A trailing variety of great beauty ; large *purple crimson* flowers ; desirable for hanging baskets........	15	298	——, **Datura, Wright's.** Flowers bell-shaped, of extraordinary size, *white* bordered with *lilac.* Two feet..	5
289	**Chelone Barbata.**........	5	299	——, **Humilis Double.** Double flowers of a rich, *golden yellow*, a magnificent, free-flowering, sweet-scented variety........	10
	A hardy perennial, about three feet high, bearing long spikes of *scarlet* bells. Flowers from July to September ; of easy culture.			**Delphinium.**	
				Flowers remarkable for their great beauty, diversity of shades and striking appearance. Hardy perennials.	

No.	PERENNIALS.	Price per P'k'ge
300	**Delphinium, Formosum.** (See Plate.) New, flowers remarkably large and brilliant; color exquisite *blue* and *white*; will flower the first season from seed. Two feet.	10
301	——, **Chinese.** Mixed. Two and a half feet.	5
302	——, **Elatum.** (Bee Larkspur.) *Blue,* two feet.......	5
303	——, **Hybridum.** Fine mixed, splendid...............	5
	Dictamus. (*Fraxinella.*)	
	Handsome, fragrant, free-flowering, herbaceous plants, suitable for mixed borders. The leaves have a very pleasant smell like lemon peel, when rubbed. The seeds frequently remain dormant for several months.	
304	**Fraxinella, Mixed.** Two feet.......................	5
305	**Erythrina or Coral Tree**	25
	This magnificent half-hardy shrub, with its *fine leaves* and most *brilliant scarlet flowers* is a great acquisition. The gorgeous spikes of scarlet flowers from one to two feet long with which it covers itself bear a resemblance to *Coral.* Although a tropical plant, it grows freely out of doors if placed in a warm situation. Cut it down to the ground before frost and protect in a cool, dry cellar during winter.	
306	**Eupatorium.** (*Fraserii.*)...........................	10
	Shrubby plants whose flowers are indispensable for bouquets. *The flowers are white, growing in graceful feathery sprays* and are admirable for mixing in with bright colored flowers.	
307	**Evening Primrose**............	5
	A well-known, showy perennial, one and a half feet high, blooming the first year from the seed.	
308	**Primrose, Hardy.** (New.) A remarkably pretty and varied strain of these popular flowers, embracing a great variety of colors.......................	20
	Feverfew. (*Matricaria.*)	
	A beautiful, half-hardy perennial, well adapted for beds.	
309	**Feverfew, Double White.** Very fine. One foot.	5
310	——, **Golden Feather.** One of the ornamental foliage plants. Very desirable for vases and baskets to mix up with other plants.....................	15
311	**Forget-me-not.** (See plate.)....................	5
	A very pretty, little, hardy perennial, about six inches high. Will thrive best in a cool, moist situation, and is well adapted for bedding or rock work.	
	Geum.	
	Handsome, free-flowering, long-blooming and remarkably showy plants. Succeeds best in a sandy loam. Hardy perennials.	
312	**Geum, Mixed**	10
	Grasses Ornamental.	
313	——, **Isolepsis Tenuila.** One of the grasses which is *a great favorite for baskets, vases, etc.* Very graceful................	25
314	——, ——, **Pampas Grass.** The most stately and magnificent ornamental grass in cultivation, producing numerous long, silken plumes of flowers. When planted on lawns the effect is very fine. Flowers the second season; require to be carefully covered during the winter, as it is not quite hardy....................	15
	Gypsophila.	
315	**Gypsophila Paniculata.** Remarkably hardy, dwarf plant, covering itself with small white flowers. Very desirable from their tenacity of life. I have known a cluster of these flowers to live three days without water and without showing signs of wilting. *Ladies will find this very desirable for ornamenting their hair, also for button-hole bouquets.*..................	5
	Hollyhocks.	
316	**Hollyhocks, English Prize.** *Very highly recommended.* The seed I have was saved from one of the finest collections in England, and is of twelve prize varieties.	15
317	——, **Tall Double Mixed.** A great improvement on the old variety. Showy perennials, four to six feet high, very effective amongst shrubbery.............	

No.	PERENNIALS.	Price per P'k'ge
318	**Lantana.** (*Fine varieties mixed.*) (See Plate.)......	15
	This showy, greenhouse plant will succeed finely in any garden soil. It forms a small bush, covering itself with *pink, yellow and orange flowers,* and also *flowers of changeable color.* Start in the house. Half-hardy	
319	**Lavender**...................................	5
	This herb I consider desirable for the flower garden from the pleasing fragrance of its leaves.	
320	**Lupins, Mixed**...............................	5
	Showy, hardy plants, two to three feet high, producing tall spikes of attractive flowers. Some species are annuals, but most of these are perennials.	
	Lychnis.	
	Very handsome and highly ornamental plants of easy culture.	
321	**Lychnis Fulgens.** Bright *scarlet,* from Siberia. One and one-half feet...........................	5
322	——, **Sieboldi.** *White,* fine; one and one-half feet.....	15
323	——, **Hybrida, mixed.** Beautiful, with large flowers varying in color from the brightest *scarlet to blood-red, purple, orange and white*........................	15
	Malva (or Mallow.)	
	Showy and desirable plants with pretty, salver-formed flowers.	
324	**Malva Minita.** Very desirable with its bright *scarlet* flowers. It blooms freely all the season..........	5
	Mimulus. (*Monkey Flower.*)	
	A half-hardy plant of the easiest culture, about nine inches high, producing a profusion of very pretty flowers. It is perennial in the greenhouse and may be easily propagated by cuttings. Select a moist, rather shaded location.	
325	**Mimulus, Cardinalis.** *Scarlet,* from California; one foot..................	5
326	——, **Hybridus.** (See Plate.) *New;* splendidly spotted and marbled in the most varied manner, rivaling the Calceolaria in the variety of its brilliant colors.......	15
327	**Mimulus, New Double.** Spotted, a beautiful variety for pot culture. Flowers *double,* of a *brilliant yellow, spotted, striped and mottled with crimson.* This, aside from its beauty, is very desirable from its remaining in bloom much longer than the single sorts.......	30
328	**Musk Plant.** (*Mimulus moschatus.*)................	5
	Much esteemed for the strong musk odor of its leaves. It has a yellow bloom. Tender perennial.	
	Nirembergia.	
	Charming little plants which flower profusely during the whole summer; exceedingly valuable for hanging baskets rustic vases and urnings; from South America.	
329	**Nirembergia, Large flowering.** A new species from the Andes. It deserves to become a general favorite both for the open garden in summer and the greenhouse in winter.............................	5
	Obeliscaria.	
	Showy plants with novel and rich colored flowers, having curious acorn-like centers; succeed in any common garden soil. From Texas.	
330	**Obeliscaria Pulcherrima.** *Fine, rich, velvety crimson* edged and tipped with yellow. One-half foot........	5
	Oxalis.	
	A splendid class of plants with richly colored flowers and dark foliage suitable for hanging pots or rustic baskets. Particularly adapted for the parlor where they bloom in mid-winter. Half-hardy perennials.	
331	**Oxalis Rosea.** *Rose* colored flowers, blooms abundantly. From Chili. One-half foot.....................	10
332	——, **Tropæoloides.** *Deep yellow flowers with brown leaves* ; a very interesting variety. One-half foot....	10
333	**Passiflora Incarnata.** This is the only *Passionflower* yet introduced that will stand our climate, requiring but a slight protection of leaves in winter. The flowers are large, nearly white, with a tripple purple and flesh colored crown......................	15
	Pentsemon.	
	Very ornamental with long and graceful spikes of *richly colored tubular flowers.* To insure bloom the first year, seed should be started early in March and planted out in May.	
334	**Pentsemon, Choice varieties mixed**............:..	20

PERENNIALS.	Price per pkg'e	PERENNIALS.	Price per pkg'e

Petunia.

Favorite, half hardy perennials, succeeding well in any rich soil. For the brilliancy and variety of their colors, their abundance of flowers, and the long duration of their blooming period, they are indispensable in any garden, and are also highly prized for growing in pots for the greenhouse or sitting room.

Petunia Hybrida Compacta Elegantissima. (New.) This new variety forms a dense globular bush of about 10 to 12 inches in height, and 14 to 15 inches in diameter, covered thickly with flowers of all colors and shades, which are peculiar to the Petunias. As a bedding plant, especially in sunny spots, this sort is unsurpassable and very effective, and it can be further recommended as a window or market plant, on account of its very regular habit and abundance of well shaped flowers.. 30

—, Vilmorin Hybrid large flowering striped. New. (For description see novelties.).................... 30
—, Hybrida Compacta Elegantissima. New....... ... 30
—, Fringed. Brilliant crimson. (New.)............. 25
—, Fringed. Satiny white, blotched with purplish crimson. 25
—, Fringed and Veined. Rose veined with black. Extremely pretty.. 25
—, Fringed. Largest flowered, mixed in great variety. 25
—, Fine Mixed,.. 5
—, Extra Choice Mixed.................................. 10
—, Buchanan's Hybrids. (See plate.) *From the finest named flowers; beautifully blotched, marbled and variegated. Flowers of extra size as well as beauty.................* 25

Pinks.

Most beautiful and highly prized, hardy perennials, growing from one to two feet high. No garden is complete without them, as they keep up a brilliant display, almost the whole season. Start early in pots, and transplant six to ten inches apart.

Pink, Carnation, Double Mixed.......................... 10
—, Carnation. *Dwarf, fiery-red. New.* Extremely double. 35
—, Picotee. These favorite plants are of great beauty, combining *the most perfect form with the richest of colors.* They have a delicate perfume, bloom profusely and are easily cultivated. The seed I send out is from one of the highly renowned growers of Germany, and is of the choicest varieties mixed.. 25
Polyanthus, Mixed. (*Primula*.).......................... 10
Showy and profuse-flowering, hardy perennials—about one foot high, blooming in April and May.
Potentilla. (See Plt.) *Extra fine, double, choice mixed.....* 20
Desirable perennials. *Flowers exceedingly brilliant and abundant.* Hardy, easy of culture, showy and *very ornamental.*
Primula Auricula. (*From Liege.*) (See Plate.)......... 25
This is *the most beautiful and desirable of the primroses,* though it has received but little attention in this country. The flower-stalk is six to eight inches high and bears a *fine truss or cluster* of from five to seven *flowers of various colors, each having a clear white or light colored eye which renders their appearance very striking.* Finest varieties mixed from named flowers. Tender perennial.

Sedum.

An exceedingly interesting and pretty little plant, growing freely on rock or rustic work, where, during the summer, it expands its brilliant, star-shaped flowers in the greatest profusion. It is very desirable for hanging baskets. Hardy perennial.
Sedum, Cœruleum. *Blue*; from Africa...................... 10

Sweet Williams.

Well-known, showy and beautiful, hardy perennials, about one foot high, making a most splendid appearance in May and June.

Sweet Williams, Mixed................................... 5
——, ——, Double Mixed.................................. 15
——, ——, Auricula Eyed................................. 10
Tritoma Uvaria, or Red Hot Poker Plant. (See Plate.)... 25
No flower excites more attention at Horticultural Fairs than this. It is a splendid evergreen perennial, producing flower stems four or five feet in height, surmounted with spikes of red and yellow flowers exceedingly striking. Admirably adapted for forming groups upon lawns or in a flower-bed, also suitable for culture in large pots. Remove the plants to the cellar in Autumn.

Wallflowers.

Very fragrant and ornamental, tender perennials, suitable for back-ground and amongst shrubbery.
—, Mixed.. 5
—, Double Mixed....................................... 15
Wallflower Harbinger. New.............................. 15
A very early flowering variety, which has produced flowers in October from seed sown in March. It is *very hardy,* and continues to produce a profusion of bright red flowers throughout the winter months. It is a decided acquisition and deserves to be widely cultivated.
Scrophularia Chrysantha................................. 15
A Perennial for decorating purposes, 1½ to 2 feet high. It forms a splendid and regular pyramid. Leaves slightly curled and of a grayish green. The flowers are round, dark red tipped with yellow. They stand in clusters distinct and free from the leaves and are very striking in effect.
Smilax... 10
There is no climbing plant in cultivation that surpasses this in beauty and grace of habit and foliage. Its cultivation has now become a specialty in every greenhouse, where it is extensively employed in all descriptions of floral decorations. When the Smilax turns yellow, it wants rest, it is not dying. Withhold water for six or eight weeks that it may rest, repot it in good soil and it will again grow.
Snap Dragon. (Anterrhinum.)............................. 5
A very showy and hardy perennial, about two feet high, and flowering well the first season. Sow the seeds early, in pots or under glass, and transplant six inches apart.

Desirable Novelties which we offer this season for the first time.

Aster, Truffaut's Fiery Scarlet........................... 25
A new very *dazzling* color, not yet existing among the tall varieties of asters.

Koeleria Berythea.. 25
An *extremely handsome* dwarf annual grass, very desirable for borders and bouquets.

Lobelia, Blue, Double..................................... 25
New and true. Fine double flowers, blooming much longer than the single ones.

Forget-me-not, White...................................... 15
New, pure white Forget-me-not, true from seed.

Petunia Grandiflora, Conpacta............................ 30
A new and generally admired Petunia, distinguished from other Petunia by its habit of growth, form of the leaves and abundance of flowers. The plants form compact bushes with a luxuriant green undulate foliage completely covered with brilliant red flowers on very short stalks. The general effect is most striking. A very recommendable novelty.

Rumex Roseus... 25
This novelty deserves to be highly recommended to every amateur who loves flowers at all. The plant forms a well-shaped compact bush of 1½ to 2 feet high. The leaves of a fine metallic green, are prettily shaped and form a flat rosette out of which rises a bush of flower stalks covered with a multitude of rose-colored flowers. The plant has also a very beautiful appearance in the Fall.

Scabiosa, Gold, or Gold Scabious......................... 20
A fine novelty, growing in richly branched bushes about a foot high with *golden-yellow leaves,* and producing numerous *scarlet and dark purple flowers,* which contrast much with the golden yellow foliage.

Verbena Venosa... 15
This *valuable acquisition* does not much resemble the common Verbena. Grows about 18 inches high, branches freely and has dark green serrated foliage. Should be sown in January and kept very moist till the seeds germinate. The quantity of flowers it produces is astonishing. It is a perpetual purple flowering plant. It does not mildew and is the proper size to contrast with most Geraniums.

ALSO OTHER NEW & DESIRABLE FLOWERS.

Astor, Coppery Scarlet, Dwarf Chrysantha Flowered.
Very fine..15

Acroclinium Atroroseum..............................10
A very large *everlasting flower* of a dark rose color.

Begonia Rex, hybrids, (ornamental leaved plants)......25
My stock of seed embraces about 30 of the most showy varieties obtained from one of the most celebrated seed growers in Europe, and may be relied upon, producing an endless variety of these most elegant plants; extra fine quality.

Browallia, Roezli......................................20
A large-flowered bushy species with *azure-blue yellow-throated* flowers. Peculiar and elegant.

Clematis, Gravolens....................................20
A free-growing hardy variety, with beautiful yellow flowers over one and one-half inches broad. Blooms from July until the middle of November. Remarkably fine.

Clematis, Pitchere.....................................20
Hardy variety of elegant habit, neat foliage and prettily shaped *brilliant scarlet* flowers. A great acquisition.

Cockscomb, Japonica Alba..............................15
A new white Japanese variety. Said to be very fine.

Columbine, Aquilegia Truncata.........................15
The California red variety elegantly variegated with orange and yellow.

Eragrostis, Brown.....................................10
A very pretty new variety of Grass. A valuable acquisition for winter bouquets for Florists and others. The panicles produce immense masses of flowers of a *reddish-brown* color and make a stiking effect.

Eschscholtzia, Mandarin...............................25
In all stages of bloom the color of this flower is wonderfully showy and lustrous. It may without much license to one's imagination be called a *Scarlet Eschscholtzia*. The inner side of the petals is of a rich orange color, the outer side of a brilliant *Scarlet*.

Ipomea, Scarlet, Ivy-leaved...........................15
An elegant climber of rapid growth, running from 6 to 10 feet in a few weeks. The leaves are of ivy shape, flowers of a very striking fiery scarlet and produced in great profusion. Most desirable.

Larkspur, Dwarf Ranunculi-Flowered....................15
The plant rises to a height of 12 to 13 inches and forms a columnar-shaped compact boquet, furnished with exceedingly numerous flower stalks, thickly studded with spikes of shining brown-violet blossoms of a uniform height. Very beautiful when grown in a mass.

Nasturtium, Spit-fire Brown...........................25
A new variety of the favorite Spit-fire differing from it by the beautiful dark brown color of its flowers.

Pansy, International Paris.............................30
A superb *strain comprising magnificent varieties*. Exhibited in the grounds of the Paris Exhibition this was selected as being the best of the numerous groups planted out, after a careful personal inspection of the Exhibition.

Pepaver Umbrosum......................................10
A very fine Poppy with flowers of a brilliant deep scarlet, marked with 4 large black spots.

Petunia, Grandiflora Superbissima Nigra...............25
Gorgeous dark red flowers, with large jet black throat. Very fine.

Petunia, Grandiflora Supurb Inimitable................25
Very robust in habit. Large rose-colored, white spotted flowers with large white, oftentimes yellow tinged throat.

Petunia, Vilmorin's Hybrid Large-flowering Striped....30
Splendid strain of very large-flowering varieties, of excellent shape and habit, specially remarkable for the rich colors and large size of their flowers, which are beautifully striped, variegated and spotted, petals nicely festooned and laciniated on the borders. Cannot be too highly recommended.

Phlox Drummondi nana Compacta Punicea.................20
This new dwarf growing Phlox is a first class novelty. The striking brilliancy of its cinnibar-scarlet is unknown in Phlox till now. The plant forms globular bushes nearly covering itself with flowers; of great value both for pot and out-door culture.

Phlox Drummondi Alba Occulata Superba.................20
Flowers in large umbels, pure *white* with *fiery red eyes*.

Phlox Drummondi Victoria..............................20
Flowers dark scarlet, more brilliant than "Fireball," very profuse in its flowers.

Stock, New Perpetual Flowering, Double White..........25
The introducer of this extremely desirable novelty says: "This variety will produce fine spikes of double white flowers from January to December; the plants grow about 12 inches high, and if plenty of room is given will grow 3 feet through and produce thousands of bunches of bloom. If sown in the Spring the plant will begin to flower in November and keep in bloom all winter and the following year, out of doors.
I gathered a large bunch of the double flowers last Christmas, frozen hard, and when put in water they opened out quite fresh, the same plants continuing to bloom throughout the year."

Xeranthemum Annun Superbissimum.......................10
The flowers of this variety are as double as those of a Double Buttercup, of globular shape, and entirely free from projectional marginal ray florets. It is said to be the finest form of Xeranthemum yet obtained. The flowers will of course be exceedingly useful for perpetual boquets.

Collections of Flower Seeds,
BY MAIL, FREE OF POSTAGE.

For the convenience of those who are unacquainted with the different varieties of flowers, or who prefer to leave the selection to us, we offer the following Collections. They contain new seed and desirable varieties, such as we recommend. Persons thus purchasing can make a great display in their flower beds, and at a much less price than when ordering by separate packages.

These Collections are always to be of our own selection, and not subject to any discount from prices given below.

COLLECTION A, contains twenty-five choice varieties of Annuals,.. $1 00
COLLECTION B, contains twelve varieties of extra fine Annuals, including choice French Asters, Double Camelia Balsams, Double German Stocks, and fine Double Zinnias... 1 00
COLLECTION C, contains ten extra choice varieties of Annuals and Perennials, embracing many of the most desirable ones in cultivation.. 1 00
COLLECTION D, contains six packets of choice, selected seeds of the finest Large Pansies, finest Carnation and Picotee Pinks, choicest Verbenas, Prize Petunias, &c... 1 00

Any one remitting $4.00 will receive the four Collections postage free.

The following additional collections will also be sent at the prices annexed, *free of postage*.

COLLECTION E, contains fifty varieties of the best Annuals, Biennials, and Perennials.............................. $2 50
COLLECTION F, contains one hundred varieties of Annuals, Biennials and Perennials, including some new and desirable sorts.. 5 00
COLLECTION G, contains ten select varieties of Greenhouse seeds... 2 00

Purchasers who prefer to make their own selections of Flower Seeds are referred to the following Prices:

The seeds will be forwarded *by mail, postpaid* to any address in the United States or Canada. on receipt of the amount of the order.

Purchasers remitting $1.00 may select Seeds, *in packets*, at Catalogue prices amounting to	$1 10
Purchasers remitting 2.00 may select Seeds, *in packets*, at Catalogue prices amounting to	2 25
Purchasers remitting 3.00 may select Seeds, *in packets*, at Catalogue prices amounting to	3 50
Purchasers remitting 4.00 may select Seeds, *in packets*, at Catalogue prices amounting to	4 75
Purchasers remitting 5.00 may select Seeds, *in packets*, at Catalogue prices amounting to	6 00
Purchasers remitting 10.00 may select Seeds, *in packets*, at Catalogue prices amounting to	12 50
Purchasers remitting 20.00 may select Seeds, *in packets*, at Catalogue prices amounting to	26 00
Purchasers remitting 30.00 may select Seeds. *in packets*, at Catalogue prices amounting to	40 00

No variation whatever will be made from the above rates. Prices to Dealers whose orders exceed the above amounts will be given upon application.

BULBS.

We this season annex to our Flower Catalogue a list of Bulbs which we offer to our patrons. We have endeavored to make a judicious selection, offering only such as are most desirable both for their beauty and adaptation to general culture. Unless otherwise specified we will send the bulbs out in September, *carefully keeping on file all orders received for them* previous to that time.

GLADIOLUS BULBS. Ready in April.

These showy flowers are very easily raised, and with their tall spikes and brilliant colors of almost every variety, simple and blended, make one of the most magnificent displays of the flower garden. By planting from May till July, a continuous succession of flowers will be secured. In planting have the rows about a foot apart, the bulbs six inches apart in the row and two or three inches below the surface. Orders will be put on file as received and filled in April.

Beautiful French Hybrid varieties—Splendid Mixtures—including nearly white, rose and crimson colors—per dozen 75
" " " " " " " per 100 by express 4 00

SPLENDID NAMED SORTS.

Adonis, *light red*	15	Imperatrice, *white, tinted blush, bright carmine, rose blotch*	15	
Archimedes, *light red, lower petals buff*	15	James Carter, *light red, with white blotch*	15	
Aristotle, *carnation rose, flecked or blotched with carmine*	15	John Bull, *yellowish white, very fine*	15	
Beronice, *rose and variegated red, with purplish carmine*	20	Laura, *orange red, with pure white blotch*	20	
Brenchleyensis, *deep scarlet, splendid for clumping*	10	Marie, *pure white, with deep carmine blotch*	30	
Celine, *rosy white ground marbled rosy carmine*	15	Mars, *beautiful scarlet*	15	
Charles Dickens, *a delicate rose striped with a darker rose, very fine*	30	Meteor, *brilliant dark red, large pure white blotch*	30	
Clemence, *satin rose feathered with bright carmine*	20	Napoleon III., *bright scarlet, the centre of the petals white striped*	15	
Don Juan, *orange fire-red, spotted with yellow*	10	Osiris, *purplish violet, white blotch*	15	
Edith, *carnation rose, with dark stripe*	15	Ophir, *dark yellow, purple blotch*	25	
Egeria, *light orange rose with with dark stripe*	15	Pegasus, *light salmon, stained with carmine and violet*	15	
Eldorado, *clear yellow, lower petals streaked red*	25	Penelope, *blush white, streaked carmine*	15	
Erullio, *white suffused with rose, brown blotch*	15	Prince Imperial, *peach blossom pink, with violet stains*	25	
Fulton, *velvety vermillion, light purple blotch*	20	Princess of Wales, *white, flaked with rosy crimson*	20	
Galathea, *blush white, carmine blotch*	15	Reine Victoria, *pure white, with carmine violet blotch*	30	
Goliath, *light red, striped with carmine*	15	Romulus, *brilliant dark red, with pure white blotch*	20	
Greuze, *intense cherry, blotche l with white*	15	Vesta, *rose white with purplish carmine blotch, on yellow ground*	25	
Ida, *white ground, flamed with carmine rose*	20	Zenobia, *rose, large white blotch, very fine*	20	

TUBEROSES. (Ready in April.)

We have a fine lot of Tuberoses from France. In planting, remove the small offsets around the main root, and plant a single tuber in a pot five or six inches wide. They should be started in April and afterwards transplanted to the open ground for summer blooming in the garden. Use good loam and leaf mould with good drainage.

First quality bulbs . . . . .15 cents each ; $1.50 per dozen.
Second quality bulbs . . . . .10 cents each; $1.00 per dozen.

HYACINTHS. (Plant in October and November.)

DOUBLE RED AND ROSE.

Alida Catherine, *deep rose, very early*	20
Bouquet Royale, *bluish pink, red eye*	25
Grootvorst, *delicate blush, very double*	25
Perruque Royale, *rose, large bells*	35
Princess Royale, *rich crimson, extra*	25

DOUBLE WHITE.

A la mode, *pink eyed, fine truss*	30
La Deesse, *white, yellow eye*	25
Nanuette, *yellow centre*	20
Sceptre d'Or, *white, orange scented*	25
Sultan Achmit, *large, very double, late*	30

DOUBLE BLUE.

Belle Mode, bright blue, beautiful	25
Duchess de Normandy, dark blue	30
Pasquin, delicate porcelain, violet eye	30

DOUBLE YELLOW.

Bouquet d'Orange, reddish yellow	30
Goethe, bright, very double, fine	30
Ophir d'Or, light yellow, fine, late	25

SINGLE RED.

Amy, bright carmine, compact truss	25
Emelina, bright rose, fine	30
Madame Hodshon, dark red, striped	25
Norma, delicate pink, large bells	25
Sultan's Favorite, rich bright rose	20

SINGLE BLUE.

Argus, deep blue, white eye	25
Blue Mourant, dark blue, black eye	25
Charles Dickens, bright blue, splendid	25
Grand Lilac, beautiful, silvery lilac, large	25
Regulus, porcelain, large truss	25

SINGLE WHITE.

Blanchard, white, purple eye	30
Hannah Moore, pure white	25
Mammoth, white, large bells	25
Queen of the Netherlands, splendid	30

SINGLE YELLOW.

Adonia, lemon yellow, good form	25
Alida Jacoba, rich, canary yellow	25
Heroine, light yellow, tipped with green	25
Rhinoceros, orange yellow	25
Victor Hugo, light orange yellow	30

Single Hyacinths, Mixed, 12 cts. each; $1.25 per doz.
Double   "   "   12 cts. each; $1.25 per doz.

JAPAN LILIES.

These superb lilies are perfectly hardy, flowers elegant and fragrant, flowering during July and August, and forming one of the principal features of the flower garden.

	Each.	Per doz.
Lilium Album, Pure White	.50	$3 00
Lilium Roseum, White, spotted with rose	.25	2 50
Lilium Auratum, Golden-rayed Japan lily	.50	5 00

AMARYLLIN—(Red Jacobean Lily).

This is always a favorite from the striking elegance of its scarlet velvet flowers. Start in the house in March and plant out in May in rich ground; roots are preserved like Dahlias during the winter. Each 25

TULIPS. (Plant in October and November.)

EARLY DWARF DUC VAN THOLL.

	Each.	Per doz.
Single Red, bordered with yellow	$ 10	21 00
Single Yellow, bright yellow	15	1 50
Single Vermilion, very bright	12	1 25
Single Gold Striped, rare, beautiful	10	1 00
Double Scarlet, bright yellow edge	5	50

SINGLE EARLY TULIPS.

Alida Maria, white and crimson	10	1 00
Alba Regalis, white, fine	10	1 00
Belle Alliance, bright scarlet	10	1 00
Bizard Proukert, yellow and red, striped	5	50
Canary Bird, rich yellow, fine cup	10	1 00
Cardinalshad, brown	10	1 00
Duc d'Orange, orange	5	50
Globe-de-Rigault, violet and white	10	1 00
Grootmeester, white, striped and feathered with scarlet	10	1 00
Lac Van Rijn, purple, white edge	5	50
Ma Plus Amiable, brown and yellow	10	1 00
Marquis de Westrade, gold yellow and red	20	2 00
Potter, violet, large flower	15	1 50
Princess of Austria, red, golden edge	10	1 00
Rachel Ruys, rosy	15	1 50
Rosa Munal, white, bordered with rose	5	50
Thomas Moore, buff orange, shaded	5	50

DOUBLE TULIPS.

Admiral Kingsbergen, yellow, with bronze stripes	5	50
Blauwe Vlag, purple blue, large	5	50
Comtesse de Pompadour, red, edged yellow	10	1 00
Duc de York, rose, white bordered	5	50

DOUBLE TULIPS.

	Each.	Per doz.
Gloria Solis, bronze, crimson border	10	1 00
Hercules, splendid striped cherry	15	1 50
La Barocque, violet, white edged	10	1 00
La Candeur, pure white, fine, early	10	1 00
Parony Gold, red and yellow	5	50
Tournesol, scarlet, yellow margined, early	10	1 00
Velvet Gem, brown velvet	15	1 50
Yellow Rose, golden yellow, very double	10	1 00

PARROT TULIPS.

Admiral of Constantinople, orange and red striped	5	50
Mixed Sorts	10	1 00
Monster Rouge	20	2 00
Perfecta, red striped	5	50
Orange	5	50

OTHER SPECIES OF TULIPS.

Cornuta (Chinese), scarlet and yellow, very curious	10	1 00
Florentina Odorata, yellow, sweet scented	10	1 00
Gesneriana, bright scarlet, fine for bedding	10	1 00
Sun's Eye, red and black	10	1 00
Persica, orange yellow, dwarf	10	1 00
Viridiflora, green, with yellow margin	10	1 00

MIXED TULIPS.

	Per doz.	Per 100.
Mixed Early Single, beautiful varieties	50	3 00
Mixed Double, very fine	50	3 00
Mixed Parrot Tulips, very showy	50	3 50

MADEIRA, OR MIGNONETTE VINE—(Hendy in April).
A beautiful climbing plant of rapid growth, adapted to out-door growth in the summer or the house in the winter. It completely covers itself with long racemes of deliciously fragrant white flowers. Plant the tuber out of doors in the Spring, and it will commence to grow at once, and if in a sheltered place, very rapidly. In the autumn cut off the tops, dig up the tubers and put them in the cellar, where they will keep as well as potatoes—or take up the bulbs and pot them for the house, where they will thrive to a wonderful extent. Tubers, each 10 cents.

SCARLET ANEMONE. (New.)
This splendid variety is almost unknown in horticulture: no plant can compete with it in beauty and brilliancy in the early spring. Flowers large and of a *dazzling vermillion*—in bloom from February to April—very valuable for bouquets. Plant bulbs in open ground in September—during winter give the plants protection with leaves. The root may remain in the ground for several years. Per bulb, 10 cts.

DOUBLE PERSIAN BUTTERCUPS—(Ranunculus).
Splendid mixed varieties. Plant in November. Each, 5 cents; per doz., 50 cents.

CROCUS.
The Crocus is a universal favorite, and, excepting the Snowdrop, is the earliest of all spring flowers, displaying its bright blossoms early in March. Plant in November.

All Colors Mixed. Per doz., 25 cents; per 100, $1.25.

SMILAX—(Ready in April).
There is no climbing plant in cultivation that surpasses this in beauty and grace of habit and foliage. Its cultivation has now become a specialty in every greenhouse, where it is extensively employed in all descriptions of floral decorations. Good bulbs, 25 cents; large bulbs, 50 cents.

LILY OF THE VALLEY. (Ready in April).
A great favorite because of its delicious color and low growth, bearing graceful bows of fairy-like bells. Each, 5 cents; per doz., 50 cents.

CROWN IMPERIALS.
An old-fashioned class of plants liked because of their highly ornamental character and early blooming. Plant in November. Mixed varieties. Each 25 cents; per doz., $3.00.

NARCISSUS.
Remarkably showy, spring-flowering bulbs, possessing a delightful fragrance. Plant in November.
Single Varieties Mixed. Per bulb, 5 cents; per doz., 50 cents.
Double Varieties Mixed. Per bulb, 10 cents; per doz., $1.00.

VARIETIES OF POTATOES.

I would advise our customers at the South to order potatoes in the fall, as there is more or less danger of their being injured by frost if forwarded between Dec. 1st and March 20th. While, therefore, I will guarantee in filling such orders to use my best judgment, all potatoes ordered to be forwarded between those dates must be at the risk of the purchaser.

CLARK'S NO. 1 POTATO.

This seedling originated in New Hampshire. It is earlier than the Early Rose and will yield from a quarter to a third more crop. It bears a close resemblance to Early Rose in appearance. It cooks mealy, is of excellent flavor, and is every way a capital variety for either the farmer or market-gardener. Raised on a large scale on my grounds the past season, I find in every instance that it surpassed the Beauty of Hebron in yield—which is saying much in favor of any sort. This potato was held in such high estimation that the entire crops of 1877 and 1878 were purchased by the Government for distribution in the South and West. 450 bushels have been raised on an acre, and 22 bushels from one peck of seed.

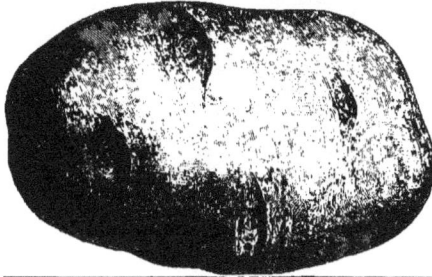

BEAUTY OF HEBRON.

Closely resembles Early Rose in shape, but is of a lighter red in color. Very prolific, being equalled in this respect, by but few, either of the early or late varieties. Quality excellent. Earlier than Early Rose, it will outyield that standard variety by from a quarter to a third. Those who think of planting Early Rose will make more by giving their seed away and thus paying double what is charged for either this or Clark's No. 1, or Ohio, if their land is strong and rather moist.

CLARK NO. 2.

This ranks with the three or four very earliest sorts before the public. It has an exceptionally smooth skin. With me it has proved but a medium cropper, while the eyes being few it don't spend well for seed purposes. Would recommend this rather to the amateur than the market gardener.

MAMMOTH PEARL.

This new Ohio seedling was selected as the best from over 2500 seedlings. It is of excellent quality for table use, large size, very handsome in appearance, and has thus far proved to be comparatively free from rot. Skin white and flesh very white; eyes few and even with the surface; in shape generally roundish; vines short and thick. In productiveness it is excelled by no potato I have ever raised.

BLISS' TRIUMPH.

I have grown this potato on my own grounds extensively this season. It is round in shape, with few eyes, and of a rich, pinkish red color. It grows to a good market size, is white fleshed and cooks mealy. In earliness it ranks with the Early Ohio, and yields about equal to it on upland.

BEAUTY OF HEBRON.

MOORE'S SEEDLING.

This is a new seeding from the State of Maine. Perhaps the very best summary of what it is, is given in the fact that Mr. Moore's neighbors paid him two dollars a bushel for all he could spare, to raise potatoes for their own use when Early Rose was selling at 35 cents per bushel. It is a seedling of Early Rose, and very closely resembles it, but grows to a larger size, yields much heavier, and for eating is most excellent, being light, dry and floury all through, while so many sorts if dry at all are so only on the outside.

LATE OHIO.

This variety was originated by Mr. Reece, the same gentleman who originated Early Ohio. It has the same excellent characteristics as that choice variety, with the addition of a characteristic vigor,—the comparative difference being just about that which is found between Early Rose and Late Rose.

THE EARLY OHIO.

This first-class potato is the first *of my own introducing* since I several years ago sent out the Excelsior. The Early Ohio is one of the numerous seedlings of the Early Rose, but while almost all of these are so like their parent as to be undistinguishable from it, the EARLY OHIO, while

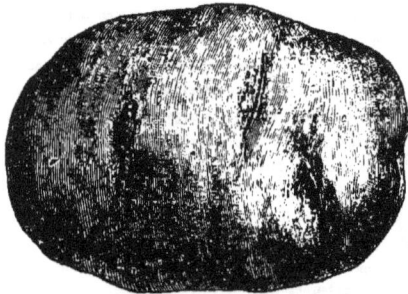

THE EARLY OHIO.

In color like the Early Rose is, in shape, distinct, being round-oblong instead of oval-oblong, so that side by side it is readily distinguishable. Eyes about as numerous as those of the parent, while the brows are rather more prominent. On the largest specimens, the cluster of eyes at the seed end are located slightly one side of the longer axis. Quality excellent. Grown side by side with the Early Rose, *it proved a week earlier, while the yield was a third greater.* To get the best results, plant on rich, rather moist soil.

"The Early Ohio Potato I got from you proved to be the best of any I have cultivated. They have three good recommendations—earliness, superior quality and an excellent cropper."
R. F. SHILLING. *Rural Dale, Ohio.*

"Early Ohio is the earliest, handsomest and best early I have ever grown, after 15 years' observation and experience—growing few small ones, and making a first class size for table use. If there is any objection, it is that they average too large. They are dry and white when cooked. As an early market potato, it almost literally drives every other out or use. When Early Rose sold at 50c. per bushel, parties would pay 50c. for Early Ohio, and declare that they would rather do it than have the Early Rose at 25c. It keeps well into the summer for an early potato. They were planted last year as late as July 10th, and made as large a crop as Early Rose planted early in May. The yield this year was 335 bushels per acre by measure." From W. S. L., in *Country Gentleman.*

Mr. TILLINGHAST, in his new work, writes as follows of the Early Ohio :—
"This is, all things considered, the best very early potato we have ever grown. It is of recent introduction. It is a seedling of the Early Rose and is named after the state in which it originated. It is of nearly the same color as its parent, but differs in shape, being more nearly round. It grows to a large size, is very productive and of first rate quality. A decided acquisition."

"Another season's trial with this splendid early potato more than confirms my previous good opinion of its excellent merits ; there is not a dissenting voice ; they all say that it is the best early potato they ever grew."
Little Sioux, Iowa.   J. L. PERKINS.

"The quality and size of your Early Ohios were excellent. Planted side by side with the Early Rose I think I can safely say that they are about a week earlier."   J. R. NICHOLS, M. D.
Haverhill, Mass.

"I obtained from Mr. Gregory, of Marblehead, last spring, some seeds of the Early Ohio. Planted side by side with the Early Rose, the Early Ohios were out of the ground first. They have proved decidedly to advantage. They were good for use, I think, ten days in advance of the Early Rose, certainly a week or more. I value them highly. I think they are of better flavor than the Early Rose.   A. G. COMINGS.
Strafford Co., N. H.

DUNMORE.

DUNMORE.

This new seedling — a white skinned and white fleshed variety, which originated in Vermont, tested in my experimental plot side by side with over forty varieties, in every requisite of a first-class potato ranks but second to the Burbank. The public will find that it is far superior in its yield, size of the tubers, their handsome appearance and fine floury quality either boiled or baked, to several of the varieties that have recently become famous. I would recommend all potato fanciers to try the Dunmore. As the same potato varies on different soils, probably on some soils the Dunmore will give greater satisfaction than the Burbank Seedling.

Mr. Tillinghast, in his book speaks as follows of the Dunmore :

"This is another valuable new variety of Mr. Gregory's introduction. It is a large, smooth, round, white potato of beautiful appearance, fine quality and enormous yielding propensities. We have grown many specimens weighing two pounds each, and have not yet seen a hollow one. Its general appearance somewhat resembles the Peerless, but it surpasses that well known variety in both yield and quality."

"From the peck of Dunmore, (the best potatoes I ever saw in my life,) I got the enormous yield of 27 bushels. I saw they were going to turn out big and I gave very large measure. I think they would weigh out over 27 bushels, and a nicer potato for table use I never saw. I would not sell my interest in them and do without them for all the other kinds raised. I took the first premium on them at the county fair and think I could have taken the same at the state fair had I taken some of them, and I should had I known they were so good at the time. We have had one of the dryest seasons known here. I honestly believe I would have had from five to eight bushels more from the peck, had they had rain when it was needed. Some of the farmers think it is impossible to raise that many from a peck of seed; they were raised on river bottom very rich soil. This statement I am ready to swear to and have the potatoes to show for themselves. They all wanted to know where I got the seed, and I told them from my old stand, of Mr Gregory. You are at liberty to make use of any part of this letter you wish to in your catalogue and if any one beats this yield, I will see what I can do with them in a good season. Some hills had 20 nice big potatoes, all large enough to eat from one eye on a piece, and one piece in a hill."   DANIEL M. CLUTE.
Fort Dodge, Webster Co., Iowa.

"From one pound of your Dunmore late potatoes, I raised eighty-five pounds. I think they are a splendid potato."
Winslow, Ill.   MRS. SARAH EBB.

"I grew 15 varieties of potatoes this year, but the Dunmore by far excelled them all in yield."   Yours Respectfully,
Mellersport, Fairfield Co., Ohio.   MATT MILLER.

BURBANK'S SEEDLING.

This, like the Early Ohio, is a seedling of Early Rose, but is of Massachusetts origin. Unlike its parent it is white skinned. It has yielded 435 bushels to the acre. Planted side by side with Early Rose, New York Late Rose, Peerless and Brownell's Beauty, it has excelled them all in yield. In beauty of form it is unexcelled, the proportions being all that can be desired, and is never hollow hearted. It has the good characteristic of yielding almost no small potatoes; while but five-sixths of the Early Rose, growing side by side, were of market size, of the Burbank forty-nine fiftieths were marketable. It has but few eyes, which are sunk but little below the surface. In quality it is fine grained, of excellent flavor either boiled or baked, is dry and floury, in fine is all that can be desired. It ranks between the very early and very late varieties. The best results have been obtained on the sandy loams of river bottoms.

In brief, the argument for sending out the new seedling is as follows: 1st, its exceptionally great productiveness; 2d, the first class quality of the potato; 3d, the capital trait for market, that it produces almost none of unmarketable size; 4th, its hardy vigor; 5th, it does not grow hollow hearted even when weighing over a pound to a single potato; 6th, the proportions and appearance are so attractive it will draw the attention of marketmen. In many sections the Burbank has become the standard late potato.

BURBANK'S SEEDLING.

"Last April 1 purchased of you one barrel of Burbank's Seedlings. Considering the dry season the yield was remarkably large. I dug 241 bushels of potatoes of superior quality." H. M. MANCHESTER. *Painesville, Ohio.*

"The Burbank's Seedling potatoes are away ahead, for yielding, of anything I ever saw. Planted by the side of others, with the same cultivation, it yielded three hundred fold more than any other kind excepting the Dunmore and Excelsior. I should have made money to have paid $20 per bushel last year, and planted all Burbank's Seedling. From the one pound you sent me I raised, without any manure or any fertilizers whatever, 187½ lbs., and not an unsizable potato in the lot. Early Rose on same ground, only gave (same number of hills) about 8 to 10 lbs., Excelsior 84, Peach Blow 43, Peerless 10 lbs." H. O. BAILEY, *Hammonds, Pa.*

"The 15 lbs. of Burbank's Seedling bought of you, yielded from the single peck 18½ bushels of large potatoes, unsurpassed in beauty and quality. No care or manuring was given them more than the other parts of the field, except they were cut finer." N. C. SNELL. *Sudbury, N. H.*

"Now as to the Burbank's Seedling, the season has been unfavorable, and the grasshoppers killed them before they were matured, but for all that, they gave good satisfaction; they realized at the rate of 420 bushels to the acre, and 98 per cent. marketable potatoes. I consider them a potato of rare excellence as a late variety." -J. L. PERKINS. *Little Sioux, Iowa.*

The Burbank Seedling was planted with all my other new sorts, and so far as a single trial is concerned, has beat them all handsomely in yield, appearance and quality, the three great essentials in a potato. *Rome, Oneida Co., N. Y.* JONATHAN TALCOTT.

I have tested over a thousand varieties of potatoes but the Burbank excels them all, growing the handsomest potatoes I ever saw, while their quality was fully equal to their looks. ANDREW LACKEY. *Haverhill, Mass.*

MR. JAS. J. H. GREGORY: Sir,—From the barrel of Burbank Seedlings you sent me I raised 133 bushels of splendid potatoes. They beat anything I ever saw. JOHN H. CARR. *Saratoga Springs, N. Y.*

WHITE ROSE POTATO.

Tuber of good size, not too large, of good market shape; quality excellent. Skin and flesh of a remarkably clear white, so that when dug they look as though made of wax. The vines are exceptionally stout and vigorous. The White Rose is a late variety and a capital cropper. Plant early in the season.

PRICE LIST OF POTATOES.

	bbl. exp.	bu'h exp.	peck exp.	25 eyes mail.	1 lb. mail.	3 lb. mail.		bbl. exp.	bu'sh exp.	peck exp.	25 eyes mail.	1 lb. mail.	3 lb. mail.
Clark's No. 1. New	5.00	2.50	1.00	50	50	1.00	Beauty of Hebron New	4.00	2.00	75	50	50	1.00
Late Ohio. New	4.50	2.25	80	50	50	1.00	Early Ohio	3.75	1.88	75	50	50	1.00
Moore's Seedling. New	5.00	2.50	1.00	50	50	1.00	Burbank's Seedling	3.75	1.88	75	50	50	1.00
Clark's No. 2. New	4.50	2.25	75	50	50	1.00	Dunmore Seedling	3.75	1.88	75	50	50	1.00
Mammoth Pearl. New	4.50	2.25	75	50	50	1.00	Extra Early Vermont	3.25	1.75	75	50	50	1.00
White Rose. New	4.00	2.00	75	50	50	1.00	Breese's No. 6, or Peerless	3.25	1.75	75	50	50	1.00
Bliss' Triumph. New	5.00	2.50	1.00	50	50	1.00	Early Rose	3.00	1.50	75	50	50	1.00

Quantity of Seed for an Acre.

Seedsmen vary much in their directions for the quantity of seed to be planted to the acre. In the following list I give the quantities of the more common sorts used by practical farmers:

Dwarf Beans, in drills	1½	bushels
Peas, that make small vines	1½	"
Peas, that make large vines	1	"
Beets, in drills	4	pounds
Cabbage, in hills	8	ounces
Cabbage, in bed to transplant	2	"
Carrots, in drills	1½	pounds
Musk Melon, in hills	1 to 1½ lbs.
Mangold Wurtzel, in drills	4 pounds
Onion, for bulbs, to sell green or to trace, in drills	6 to 8 lbs.
Onion, for dry bulbs, in drills	4½ pounds
Onion seed for Setts, in drills	30 pounds
Onion Setts, in drills	10 bushels
Potatoes, in drills, cut, depends on number of eyes	8 to 14 do.
Radish, in drills	5 pounds
Spinach, in drills	10 to 15 lbs
Sage, in drills	4 to 6 lbs
Squash, running varieties, in hills	2 to 2½ lbs
Tomato, in bed to transplant	2 ounces
Turnip, in drills	1 to 1½ lbs

CHINESE FODDER PLANT.

I have grown this for two years, having obtained my seed from the Chinese department of the Centennial Exhibition, after its close. It belongs to the list of forage plants which are now creating such an interest among agriculturists. I regret that I did not select a specimen for engraving. It starts like a slender corn, and pushing rapidly along attains to the height of five feet under ordinary corn cultivation. The stalks are stout and throw off a great abundance of very broad leaves, nearly as broad as the smaller varieties of corn. It throws out several stalks from a single seed, spindles closely resembling that of corn, on which the seed grows most abundantly, yielding a crop of from thirty to forty or more bushels to the acre. The seed ripens readily as far north as central New England. I find fowl and other animals very fond of them. The seeds resemble millet, but are larger and do not grow as in the common varieties. When raised for fodder this plant should be cut before the seed matures, as otherwise the stalk becomes woody. Per package 15 cts.; per pint, by mail, $1.00.

BLUNT'S PROLIFIC FIELD CORN.

Prof. Blunt has produced this remarkably prolific field corn (*it will yield all the way from three to six ears to the stalk*) by proceeding on the theory that since every joint that has a groove contains an elementary ear, therefore, by a scientific selection of seed ears and proper cultivation, these dormant ears may be fully developed. The result is that, starting with two ears to the stalk he has succeeded in so improving the variety, the yield now is as stated above, from three to six ears to each stalk. It has cropped over a hundred bushels shelled corn to the acre. My customers in New England will please note that *the corn is too late to mature in their section.*

For prices, see page 22.

(Reduced from Rural New Yorker.)

DOURA.

This is a forage and grain plant about the value of which the public has been somewhat confused, from a general ignorance of the fact that the varieties had characteristics very distinct from each other, one of them growing from six to sixteen stalks from a single seed, while two varieties grow but a single stalk. Says the careful editor of the Rural New Yorker, writing of the variety which yields several stalks from one seed, after a careful test of the various plants for fodder, "we believe it will prove of great value. We like it better than any fodder plant we have ever tried." It stands firm against the highest wind, grows to the height of nine or ten feet, will make a second growth after cutting. It is allied to sugar cane, is much sweeter than corn stalk, and cattle and horses eat it ravenously. From six to sixteen stalks grow from a single seed. The varieties of Doura, known as China corn and Egyptian corn, grow but a single stalk from a single seed, but being much earlier will ripen their seed in the north, giving from 30 to 60 bushels to the acre. "In California," a correspondent writes me, "it will ripen a crop in about 78 days from time of planting, yielding from 30 to 60 bushels of pure grain per acre. Every thirty days thereafter until frost, it will make a crop of heads about half as great as the first crop. This second crop grows on limbs which start from the joints or from suckers which start from the roots. For buns, puddings, &c., the China corn only is used, the Egyptian being grown only for cattle feed. It must be ground *very fine* and *bolted*, then used the same as wheat flour. Cooked whole, as rice, it is also excellent." My correspondent, in whose family it has been grown for two generations, states that "he did not realize its value until he noted that when during a severe drought all the crops failed, this was not affected in the least." Price per mail, postpaid, per package, 10 cents ; per ¼ lb., 20 cents ; per lb., 60 cents. Four pounds will plant an acre.

CUZCO CORN.

This is the giant corn of South America. The kernels measure one inch long by ⅜ inch wide. To realize what this means, let any of my customers measure kernels from their largest varieties. I send this out purely as experimental, having had no personal experience with it. Such a giant would doubtless grow an immense stalk in any part of our country, but could not be expected to mature seed in the north. These monster seeds are unique curiosities. Per package, 15 cents.

Some Remarks and Advice about Peas.

HANCOCK EARLY. Tested side by side, with fifteen of the earliest varieties, this proved in purity, earliness, productiveness and quality, to be one of the *very best.*

HAIR'S DWARF MAMMOTH. Peas and pods very large ; a wrinkled variety, popular for the family garden.

CARTER'S FIRST CROP is earlier than DAN O'ROURKE, but the pods are smaller. A further trial, by market gardeners, has brought the CARACTACUS into high favor. The pods are of a good size for an early pea and well filled. MCLEAN'S ADVANCER is a *first rate* second-early for market or family use ; pods large, well-filled and numerous. One of the sweet, wrinkled class.

BROWN'S EARLY DWARF MARROWFAT PEA will be found to be the earliest and most dwarf of all Marrowfats. This is one of the very few varieties of American origin.

ALPHA. This dwarf wrinkled pea in yield probably surpasses any of the first only sorts ; it is distinguished for earliness, productiveness and sweetness. A decided acquisition. Market Gardeners are much pleased with this as a first early.

Quantities of seed required for a given length of drill.

This table is probably as correct as such general statements can be made.

Asparagus 1 oz. to 60 ft. of drill	Parsley... 1 oz. to 150 ft. of drill
Beet ......1 oz. to 50 "	Parsnip...1 oz. to 200 "
Beans dwf. 1 qt. to 200 "	Peas...1 qt. 100 to 150 "
Carrot.....1 oz. to 150 ft. of drill	Radish.....1 oz. to 100 ft. of drill
Endive.....1 oz. to 150 "	Salsify....1 oz. to 70 "
Okra.......1 oz. to 40 "	Spinach...1 oz. to 160 "
Onion......1 oz. to 100 "	Turnip....1 oz. to 150 "
Onion Sets 1 qt. to 20 "	

Weight of Grass, Clover and Grain Seeds and Potatoes per bushel

(see page 31). Pecks 1-4 of bushel weight.

Lawn Grass..................14 lbs	White Russian Spring Wheat, 60 lbs
Timothy or Herd Grass.....45 "	Champlain Wheat ....60 "
Red Top Grass...........10 "	Defiance Wheat.............60 "
Orchard Grass............14 "	Hulless Barley.............48 "
Hungarian ....48 "	Silver Hull Buckwheat......48 "
German or Golden Millet...48 "	White Zealand Oats........32 "
Pearl Millet..............48 "	Chinese Hulless Oats.......32 "
Red Clover................60 "	Probstier Oats............32 "
Alsike Clover.............60 "	Potatoes.................60 "

WHITE ZEALAND OATS.

I present my customers with an engraving of the new oat, made from a photograph of a couple of average heads grown on my experimental grounds this season. In going over the field, note-book in hand, I found that while every variety of oats had rusted badly (they were all planted rather late), and had for the most part, fallen down, there was one kind standing up very conspicuously with scarcely the sign of any rust. The straw of it was extraordinarily tall and stout, and the leaves remarkably broad. The heads were 15 inches in length and well filled. On turning to my note-book I found this new sort was the White Zealand Oats. Taken altogether, its superior merits were so striking, that I believe it well worthy of an introduction among my patrons. Price by bushel, by express, or freight, $1.75; per peck, 60 cts.; 1 lb. by mail, 35 cts.; 3 lbs., £0.90.

HULLESS BARLEY.

This grain is sometimes called by the name of "Barley Wheat." The gentleman who first called my attention to it states that in 1868 he selected a single head from a sack containing an admixture of this with the common variety of Barley, and after an experience of several years, he is satisfied that he has an acquisition.

The grass is entirely free from hulls, shelling out from the head, bright and clear, as though artificially hulled. A correspondent writes me: "Our horses and cows are very fond of it, liking it better than the common kind. I have grown it side by side with the common barley, and it stands up equally well, though I should like it better if it had a little stiffer stalk." It is very obvious, a hulless barley must prove of greater value for all uses within the family, than the common kind. The grain is large, plump and handsome.

The gentleman from whom I received the Hulless Barley informs me that he has at times raised two crops the same season. Ripening by July 4, there was ample time to raise a second crop from seed of the growth of the same season.

Per package, 15 cts.; per pound, 50 cts.; 3 pounds, $1.00; per peck by express or freight, $3.00; per bushel, $10.00.

WHITE PROBSTEIER OATS. These are a German oat, well adapted to this climate, and so far have not shown that inclination to deteriorate or "run out" that is usually exhibited by other heavy varieties. It is somewhat taller than the common variety, of strong rank growth, the leaves being very long and wide, and of an unusually dark green color. The straw is *Coarse and Strong and not liable to lodge*. This is also a bush Oat, the grain being distributed on all sides of the heads, which are large and well filled. The kernels are large and plump, and enveloped in a *soft, thin, white husk*. It ripens two or three days later and yields much better than the common variety. The yield has varied from 56 bushels to 98 bushels per acre; the average for six years being a little over 74 bushels per acre. They sometimes weigh 39 lbs. to the measured bushel. Price per bushel (of 32 lbs.), $1.25; half bushels, .75; peck, .50.

PRICES OF CRANBERRY PLANTS ROOTED.

I have arranged with a reliable grower to supply Cranberry Plants at the following prices:—

10,000 plants by Express, freight paid by purchaser, sufficient
for one acre at two feet apart $25 00
         If sent by mail prepaid by me 30 00
5,000 . . . .   "   "   " 15 00
1,000 . . . .   "   "   " 3 00
100 . . . .   "   "   " 50

Mansfield Creeper, a new upland variety, habit and growth different from other varieties—these are furnished by cuttings, or shoots—take root freely, and are as safe in planting as rooted varieties . . . . . Price per 100 35

Mansfield Creeper Full directions for cultivation sent with each lot ordered. No plants sent C. O. D.

Eaton Black Bell Cranberry. Berries are not very large, but uniform in size, and of dark color. The plant is very productive. It ripens by the 5th of September, *two or three weeks earlier than other varieties*, which gives them a higher price in the market. Plants furnished by the 100 or 1,000. Price per 100, per mail, 55 cts. Per 1000, $4.00.

Bell Cranberry.

TESTIMONIALS.

My friends are oftentimes pleased, without any solicitation of mine, to write me the results of their trials of my seed. Here are extracts from a few which I have taken the liberty to publish. *They are from forty-two different States and Territories.*

"I received some Burbank Seedling Potatoes two years ago and they have proved to be the best Potatoes I ever raised. I sent Potatoes to our fair last Fall, one year old, perfectly sound,—in fact, they seem to never rot." D. S. CLEMENT, *Marysville, Ohio.*

"After three years' trial of your Danvers Carrots I can say that they surpass any ever seen here." JOHN TEACKLE, *Baltimore, Maryland.*

"I raised the largest and best Cauliflower in this neighborhood, last year, from seeds of the Early Paris variety purchased of you." H. H. EVERTON, *Monroe, Ill.*

"Some of the Short Horn Carrots from your seeds weighed two pounds." MRS. SIDNEY MORSE, *Leavenworth, Kansas.*

"The Egyptian Sweet Corn which I raised last year from seed obtained from you, surpasses anything in sweetness I ever raised. It is perfectly luscious." C. M. HARRISON, *East Orange, N. J.*

"I planted your Canada Victor Tomato seed in April; on the 9th of July I picked ripe tomatoes from the vines for dinner." Miss MARY MAHSTON, *Gardiner, Me.*

"The package of Mammoth Cabbage seed I purchased of you last spring, are producing the finest lot of Cabbage in this part of the country." W. H. MAYFIELD, *Hewietta, Ky.*

"Your Early Red Globe Onion grew a larger crop than any other kind I have grown. I had them to weigh from 18 to 22 ounces." JOHN WINDROSS, *Pensaukee, Wis.*

"I have gardened in this country for 17 years, and have tried almost every firm in the United States. Have had the best success with your seeds. I marketed 4 tons of Marblehead Cabbage, last fall, that averaged 34 pounds apiece, being trimmed close. Fotler's weighing 30 pounds each. Red Globe Onion, 52 weighing 62 1-2 pounds. Yellow Danvers, 54 weighed 61 3-4 pounds." JOHN A. STROUSE, *Morrison, Col.*

"The seeds I purchased of you last spring produced fine crops of the first quality. Canada Victor Tomato ripened two weeks earlier than any in my neighborhood." N. G. DAVIS, *Newmarket, N. H.*

"Your Canada Victor Tomato gave great satisfaction. I had ripe tomatoes two weeks earlier than any of my neighbors." MARY WILSON, *Clifton Mills, West Va.*

"Last season I raised from one hill, 51 Cocoanut Squashes, and for quality they were superior to anything I ever saw in the Squash line, for fall use." DR. J. H. WESTCOTT, *Norwich, N. Y.*

"I had some Paragon Tomatoes from your seeds, that measured 12 1-4 inches in circumference, and 3 inches in thickness, smooth as an apple. Very few seed in them—very rich and fine flavor, and they ripened all over at the same time." A. G. RAMSEY, *Warrensburg, Mo.*

"I have had my seed of you for the last fifteen years, and have found them true to name and quality." P. SYKES, *Westfield Centre, Minn.*

"Your Orange Jelly Turnips are a splendid Turnip and stand the winter well." JAMES E. FORD, *Shady Grove, La.*

"Your seeds I have planted for many years, and they have given us satisfaction, always. Your care and enterprise in introducing all that is new and valuable, has made you a name in the Great West, above all others, as a seedsman, and we have much to thank you for." THOMAS WARDALL, *St. Ansgar, Iowa.*

"From the peck of Burbank's Seedlings I got of you in the spring of 1877, I raised 47 bushels at as nice potatoes as I ever saw, and in 1878 I raised at the rate of 450 bushels per acre." JOSEPH JOHNSON, *Kokomo, Indiana.*

"Your Cabbages and Onions astonish the inhabitants. No one would believe that I grew the onions from seed, as they have made so many fruitless attempts to raise them in this section. I was never better satisfied." J. S. STEBBINS, *Riceboro, Ga.*

"Old Pete says you are the best seedsman in this world's paradise. The old man lives with Samuel Purchase, in the town of Olive, and his equal for gardening is not found in our whole county, both as for quantity and quality; and the old man says the whole secret is, he gets his seeds from you." MRS. VICTORINE HICKMAN, *Grand Haven City, Mich.*

"We are happy to say we have always found your seeds perfectly reliable. Never in a single instance have they failed to germinate or be true to their kind." S. W. NASH, *Wallingford (Conn.) Community.*

"I have had sufficient experience with your seeds to know they can be relied on, which I can't well say of others." L. L. C. ELLIOTT, *Camden, Arkansas.*

"Your Hubbard Squash, Phinney's Watermelon and Trophy Tomato are ahead of anything we ever had here." NEPTUNE LYNCH, *Horse Plains, Montana Territory.*

"Last season I raised a very good patch of cabbage, about two tons of Fotler's Early Drumhead. I weighed several heads after taking away the loose leaves, and found many which weighed 33 lbs." REES K. LEWELLYN, P. M., *Fountain, Utah.*

"I obtained some Sandringham Celery seed from you last year, and I found it far superior to any I ever saw for early marketing." PAUL M. BARKER, *Newport, R. I.*

"I have used your garden seeds and they are the best that I have ever used. *I have raised thirty tons of the Mammoth Cabbage to the acre.*" A. J. BARRETT, *Dayton, Nevada.*

"I think it would be a difficult matter to find a finer lot of Cabbages than those I have growing from Little Pixie, Cannon Ball and Winnigstadt seed obtained from you in the winter." W. S. HARLEY, *Walterboro, S. C.*

"Your seeds give me great satisfaction. I consider it cheaper to buy of you than to raise my own." J. RAINS, *Washington, Idaho.*

"I have grown Sill's Hybrid Muskmelon for two years past, and would say that the quality of the fruit is truly *delicious.*" GEO. W. STETSON, *Barre, Mass.*

"Your Yellow Danvers Onion seed is the best I have ever planted. Your Marblehead Mammoth Cabbages do very well here. I have had several heads that weighed 33 lbs." ALONZO FORBES, *Jolon, Cal.*

"I have for the past three years sent to you for seeds and have always found them what Andrew Jackson would call O. K." G. W. CATE, *N. Montpelier, Vt.*

"The California Mammoth White Radish was very nice. There were one or two that were 18 inches long and 5 1-4 inches through." WM. H. TAYLOR, *Barnes, Penn.*

"I had Gen. Grant Tomato seed from you last season, and find them the best of any yet tried for this climate." C. P. ROGERS, *Frederica, Del.*

"I planted your Canada Victor seed after my other tomatoes were up and had about eight leaves on them, and the Victor beat them getting ripe by two weeks." LUCY ROBINSON, *Oregon.*

"Your seed were as usual *first rate*. I have the finest field of Cabbages raised in this section, Winnigstadt, Premium Flat Dutch and Stone Mason." J. P. JANES, *Jacksonville, Fla.*

"The seed I bought from you last spring were the best lot of seed I ever bought. They were 'all right.'" GEO. S. POWELL, *Catawba, N. C.*

"The seed which I procured in the spring have given entire satisfaction, especially the Lettuce and Onions and Cabbage." REV. JOHN H. RICE, *Memphis, Tenn.*

"Your seed do better in this section than any other. I speak knowingly, as I have tested a great many during the past few years." C. P. ELGIN, *Corinth, Miss.*

"Your Marblehead cabbage seed, purchased from your house last season, proved to be the best in this section of the country." HENRY HOWARD, *Walla Walla, Wash. Ter.*

"The Turnip seeds I bought of you proved to be of fine quality." JAMES M. CONNAWAY, *Rockford, Alabama.*

"I sent to you for seed in 1860, and I sent again in 1869, and received good seeds." P. W. WEBB, *Tecumseh, Nebraska.*

"Your seeds are just what you represent in every instance." W. R. PRICE, *Courtney, Texas.*

☞ If any of my friends wishing for Circulars to distribute to their neighbors, will write me to that effect, I will send extra copies free. ☜

www.ingramcontent.com/pod-product-compliance
Lightning Source LLC
Chambersburg PA
CBHW022012190326
41519CB00010B/1484